Material Recovery Facility Design Manual

CalRecovery
and
PEER Consultants

C. K. SMOLEY

Library of Congress Cataloging-in-Publication Data

Material recovery facility design manual / PEER Consultants and
    CalRecovery, Inc.
        p.   cm.
    Includes bibliographical references.
    ISBN 0-87371-944-1 (acid-free paper)
    1. Recycling (Waste, etc.)—Handbooks, manuals, etc.   I. PEER
Consultants.   II. CalRecovery, Inc.
TD794.5.M39   1993
628.4′458—dc20   92-34232
                                                      CIP

Direct all inquiries to CRC Press, Inc., 2000 Corporate Blvd., N.W.,
Boca Raton, Florida 33431.

PRINTED IN THE UNITED STATES OF AMERICA
1 2 3 4 5 6 7 8 9 0
Printed on acid-free paper

Preface

The purpose of this publication is to address the technical and economic aspects of material recovery facility (MRF) equipment and technology in a way that will be helpful to solid waste planners and engineers at the local community level. This book points out what technically can be done, what material specifications can be achieved, and what different manual and mechanical materials separation and recovery approaches can cost.

This handbook is designed for use primarily by engineering or other technically trained personnel who are engaged in some aspect of design, specification, purchase, or implementation of MRFs. Sources of information for this document included the design engineering community, vendors of equipment, and various studies funded by the U.S. Environmental Protection Agency. Of course, many presently operating MRFs also served as prime sources of information.

Notice

This manual is intended to assist regional, state, and local community personnel, as well as individuals or corporations considering the establishment of a material recovery facility (MRF). This document is not a regulation and should not be used as such. The users of this handbook must exercise their discretion in using the information contained herein as well as other relevant information when evaluation MRFs. The development and compilation of the guidance and information contained in this handbook had been funded wholly or in part by the United States Environmental Protection Agency through PEER Consultants, P.C.

Mention of trade names or commercial products does not constitute endorsement or recommendation for use.

Acknowledgment

This publication was prepared by PEER Consultants, P.C., under sponsorship of the U.S. Environment Protection Agency. Edwin Barth of the Center for Environmental Research Information, Cincinnati, Ohio, was the Technical Project Manager responsible for the preparation of this document. Steven J. Levy of the Office of Solid Waste and Lynnann Hitchins of the Risk Reduction Engineering Laboratory provided guidance and support. Special acknowledgment is given to:

- Darlene Snow and Jack Legler, National Solid Waste Management Association;

- James Meszaros and Robert Davis, Browning Ferris Industries;

- Jessie Buggs, Prince Georges County, MD, Residential Programs;

- Chaz Miller, The Glass Packaging Institute;

- Richard Kattar, New England, CRInc.;

- Bill Moore and Dan Kemna, Waste Management, Inc.;

- Mike McCullough, Ohio Environmental Protection Agency,

all of whom served as advisory committee members and/or technical reviewers and contributors.

Participating in the development of this book were Joseph T. Swartzbaugh and Donovan S. Duvall of PEER Consultants and Luis F. Diaz and George M. Savage of CalRecovery, Inc.

Table of Contents

Figures

Tables

Material Recovery Facility Design Manual

CHAPTER 1

Introduction

The purpose of this document is to address the technical and economic aspects of material recovery facility (MRF) equipment and technology in such a manner that the document may be of assistance to solid waste planners and engineers at the local community level. This document points out what technically can be done, what material specifications can be achieved, and what the different manual and mechanical materials separation and recovery approaches can cost.

This technology transfer document is a handbook intended for use primarily (but not exclusively) by engineering or other technically trained personnel who are engaged in some aspect of specification, purchase, or implementation of MRFs. It should be noted that this handbook offers some design-related information, but is not intended to be a design guide. For this document, a MRF is defined as a central operation where commingled and/or source separated recyclables are processed mechanically or manually. Here, a separation and/or beneficiation of recyclables prepares them to meet market specifications for sale. Sources of information for this document include the design engineering community, vendors of equipment, and U.S. Environmental Protection Agency (U.S. EPA) and other federal agency documentation of process evaluations for MRFs. Of course, many presently operating MRFs also served as prime sources of information.

This document focuses primarily on equipment and methods for the separation and handling of separable or already source-separated, recyclable constituents in the typical municipal solid waste stream. For any single recyclable constituent within the solid waste stream, alternative approaches are identified for separation and recovery (namely, manual versus fully automated versus some kind of combined approach utilizing both manual and mechanical methods). For each piece of equipment in any approach, the document addresses: the basis of design; theory of operation; sizing; and equipment needs such as shredders, balers, etc. The document descriptions include any limitations on materials in the feed to the equipment, area requirements, building requirements, possible citing and permitting requirements, industrial health concerns, and level of operator experience and training needed for proper operation. In addition, economic factors are discussed: purchase price; utility requirements; maintenance costs; labor costs; auxiliary equipment purchase needs; sizing; space requirements; redundancy requirements; and all aspects necessary for the development of performance and equipment specifications. The key focus is on the percent waste reduction (efficiency) and costs. The document is intended to give guidance to the engineer as to what should be incorporated into startup, implementation and acceptance testing of any equipment and systems to be included in the MRF.

MRFs are relatively new in the solid waste management field, but their popularity is fast increasing. In the early 1980s, the first MRF was established in Groton, Connecticut. This facility was primitive by today's standards (not full scale). The Groton facility today is operational, but it does not accept any of the paper or plastic streams, which are vital and integral components of any full-scale MRF. Recently 104 MRFs were identified in the U.S. with about one-third operational, about half (51 percent) in early or advanced stages of planning, 11 percent under construction, and 4 MRFs temporarily shutdown or undergoing significant retrofitting (GAA, 1990).

An obvious question to many parties is the sudden increased interest in the MRF as an approach to processing solid waste. Interest stems from:

- desire to reduce MSW going to landfill;
- achieve this reduction by maximizing recycling; and
- MRFs simplify generator requirements, and thus increase participation in recycling.

The probable principal reason is that as solid waste disposal costs keep rising, a greater impetus develops in the favor of recycling, and the development of more MRFs. For example, when landfill costs were less than $10 per ton, recycling most of the waste stream was not economically attractive to the waste industry.   However, now with tipping fees in some areas approaching, or even exceeding, $100 per ton, waste managers are willing to spend more time and money on recycling.

The appeal of MRFs seems to fall into three principal categories (Biocycle, 1990):

- the feedstock of most MRFs is commingled recyclables;
- collection needs can be simplified, and
- materials processed through MRFs are more marketable.

Citizens are encouraged to participate in a MRF operation, and as a result of this success of citizen participation, higher volumes of materials will be taken from the solid waste stream.   Second, because of the commingled nature of recyclables, collection vehicle needs can be simplified.   Need for multicompartment vehicles is reduced; normally only two compartments are required. Collection times and costs can also be reduced, because less time is spent at the curb sorting materials or emptying several containers.   Finally, most advocates believe that materials processed through MRFs are more marketable. They feel that the products of MRFs are cleaner, can better meet industry standards, and that the consistent volume of material that they can generate helps to assure a market.

The design of MRFs must be such that commingled recyclables can be separated, and the separated materials processed into marketable commodities. Most MRF vendors have their own basic design concept, but they maintain the flexibility to modify their design depending upon the specific requirements of the individual MRF.   In other words, vendors can respond to the needs of the community and try to provide a system that will process the recyclables that are common to that community.

Even though many of the MRF systems are highly mechanized, there are still many jobs that are best done by humans. For example, nearly all of the systems presently in use hand sort glass by color.   This approach is still the most reliable way to ensure quality. In any case, the trade-off between the manual and mechanical MRFs is capital cost versus operating cost. The highly mechanical systems have a capital cost that ranges from 75 to 100 percent higher than those for the manual systems. A life-cycle cost analysis over the operational life of the facility may show that the higher operational costs for labor intensive manual systems will become more important than the initial higher capital costs for mechanical systems.   However, operating experience for either type system is still too limited to allow independent evaluation of the actual useful life of such facilities.

The chief processing problem in any MRF is separating the mixed bottles and cans. Most of the MRF systems utilize a magnet to pull the steel cans from the mixed materials. Once the ferrous material is separated it can be either shredded or baled, depending upon the market. The remaining fraction then includes the glass, aluminum, and plastics. At this point in the process, mechanical systems can be used to either separate the lighter fraction, aluminum and plastics, from the glass. The more manual systems normally utilize workers to perform this function as well as to separate the glass by colors. The mechanical systems, however, still normally use a manual sort for separating glass colors; after the glass is separated it can be crushed and stored for market. Aluminum can be separated from the mixed materials either manually or with aluminum separating equipment such as eddy current separators. When plastics are accepted at the MRF, they are normally separated by type.

While theoretically all the materials coming into a MRF should be recyclable, it has been shown that systems always have some residues. Such residues include contamination that is mixed in with the recyclables, some nonrecoverable materials (such as broken mixed glass in a commingled, source-separated stream) and some materials which cannot be properly recognized by the sort mechanism used in the MRF. The amount of residue depends heavily upon the processing efficiency of the facilities, and this is governed in many instances by how well the community has separated its recyclables, and by what collection method is used. For example, if residents persist in disposing of nonrecyclable material in the system, then understandably the amount of residue increases.

CHAPTER 2

Specific Approaches to
Materials Recovery

2.1 INTRODUCTION

The purpose of this Section is to provide the reader with a basis for understanding, comparing, and evaluating the relative merits of different approaches and solutions to the problems of materials recovery. The information presented herein should be of value to those public officials charged with review and decision making responsibilities as well as to those individuals responsible for materials recovery facility planning and design.

2.2 CONCEPT DESCRIPTION

Several approaches have been proposed for the recovery of materials from municipal wastes over the past 40 years. The proposed schemes range from low technology (i.e., low capacity, relatively simple, labor-intensive, minimum hardware) processes to high technology (i.e., high capacity, relatively complex, mechanical-intensive, high capital and operation and maintenance costs) concepts. Similarly, a myriad of devices have been suggested for segregating one or more materials from the waste stream. The decision to select one approach versus another one is affected by a number of factors. Some of these factors include: size, cost, location, environmental impacts, and economic conditions of the particular area. There are a large number of conceptual designs and combinations of equipment that could be described, designed, and implemented. The scope of this Manual does not permit lengthy descriptions of all possibilities. Consequently, the specific approaches and concepts described in this Manual have been limited to only some options. In addition, in order to avoid misunderstandings, definitions of important terms used in the Manual are provided in Appendix 1. Descriptions of the concepts and conditions used in the Manual follow.

2.2.1 Basic Materials Recovery Facility (MRF)

The discussion of design and operating procedures for MRFs is based on a "conceptual" or "basic" MRF. Some variations from the basic design are used to focus on some specific points.

The basic MRF discussed in this Manual is one which is designed, constructed, and operated under a few sets of conditions. The conditions apply to the incoming waste and to the storage, processing, and shipping of the recovered materials. The conditions for each of these items are as follows:

1) Incoming Waste

   • The Facility receives only source separated materials. The materials are delivered in two distinct forms. One stream consists of paper and the other of commingled containers.

   • Materials that would be delivered to the Facility in the commingled container stream include: ferrous metals, aluminum, glass, PET, and HDPE.

   • Recyclable materials are delivered to the facility via commercial collection vehicles.

   • The Facility is not designed to accommodate self-haul vehicles.

2) Storage, Processing, and Shipping

   • For each of the two incoming fractions of recyclable materials, the Facility will provide raw materials storage, means for separation and processing, storage for finished products, and means for shipping the finished products in the most appropriate form. A description of these conditions is presented in Figure 2-1.

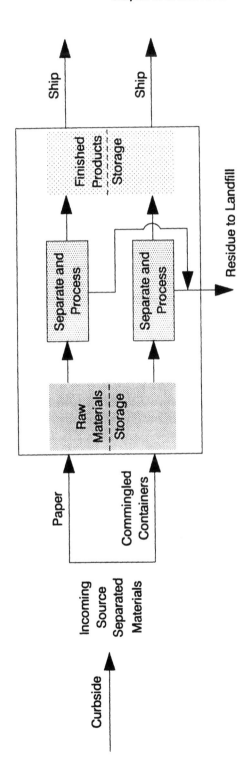

Figure 2-1. Flow chart source separated materials.

- The types of finished products from the incoming paper stream include: newspaper, mixed paper, and some corrugated.

- The forms and conditions in which finished products are to be shipped (and thus, the processing necessary to produce those forms and to prepare them for shipment) are dependent upon the economics of processing as well as upon the specifications of the markets. Financial viability and market requirements are two major considerations that affect the design of a MRF. It would not be financially feasible, for example, to increase product quality beyond that which would be necessary to market the entire output from the Facility.

Currently markets do not exist for all of the materials that can be recovered from the waste stream. In those markets that do exist, the market specifications are varied. Some of the forms and conditions applicable to the finished products include:

- Paper
  - Separated by grade
  - Baled (bale size and/or weight specified), or loose
  - Dry (or may include wet)
  - Clean (or contaminated or not weathered)

- Ferrous Containers
  - Flattened, unflattened, or shredded
  - Labels removed (or not)
  - Clean (or with limited food contamination)
  - May or may not include bimetal
  - Loose, baled, or densified into biscuit form with bale or biscuit size and/or weight specified

- Aluminum Containers
  - Flattened, shredded, baled, or densified into biscuit form with bale or biscuit size and/or weight specified
  - Free of moisture, dirt, steel, foil, lead, plastic, glass, wood, grease, oil, or other foreign substances

- PET/HDPE
  - Baled, granulated, or loose
  - Separated by color or type or mixed
  - With or without caps

- Glass
  - Separated by color and/or mixed
  - Size of cullet (maximum and/or minimum) specified
  - Nature and amount of allowable contamination, if any, specified

- General
  - Available markets for secondary materials typically specify the means of packaging and shipping each product. The specifications depend upon location and end-use. The specifications often include the following:

    o Skids or pallets
    o Bundles, bins, boxes, cartons, or drums
    o Trailer loads
    o Roll-offs
    o Rail cars

2.2.2 Variations from the Basic MRF

In the text to follow, references and comments will be made to designs of MRFs which vary from the "basic" MRF. The comments are made because there are some facilities that have been designed in that manner. In addition, the current climate in the industry points in those directions. These variations include:

1) The Facility receives only source separated materials in a single incoming waste stream.

2) The Facility also receives other source separated materials such as yard waste, wood waste, tires, corrugated, mixed metals, used motor oil, lead and batteries, used clothing, appliances, other plastics, etc.

3) The Facility receives mixed municipal solid waste (MSW) in addition to source separated materials.

4) The Facility receives only mixed MSW.

5) In addition to receiving recyclables and/or mixed MSW from commercial haulers, the Facility receives mixed MSW from self-haul vehicles.

6) In addition to the recovery of recyclables, the Facility produces a Refuse Derived Fuel (RDF).

7) In addition to the recovery of recyclables, the Facility prepares a compostable feedstock.

2.3 TECHNICAL

2.3.1 Waste Characterization

In order to properly design a MRF it is advisable (in some states, necessary to comply with legislation), among other tasks, to perform an analysis of the waste stream, i.e., a waste characterization study, so that the variety and relative quantities of incoming materials can be identified and determined. Actual field measurements are the preferred method of waste analysis. An example of the composition of residential curbside recyclables is presented in Table 2-1. These relative quantities are influenced, for each community, by various factors including:

1) State "bottle bills" which offer a financial incentive to the consumer to return the container (metal, glass, and/or plastic) to the seller thereby reducing the quantity of the item(s) from the incoming waste stream.

2) Community demographics which, as a result of its urban, suburban or rural nature, income level, reading habits, residential/commercial/industrial mix, and other factors increase or decrease the relative quantity of any item(s) in the incoming waste stream.

Table 2-1. Example of Composition of Residential Curbside Recyclables*

Material	% by weight
Newspaper	33
Mixed Paper	41
Total Paper	**74**
Glass Bottles	
Clear	11
Green	4
Brown	4
Tin Cans	4
Aluminum Cans	1
Plastic Containers	
PET	1
HDPE	1
Total Commingled Containers	**26**
TOTAL	**100**

* Not to be considered as either average or typical

3) Significant variations in the relative quantity of an item(s) in the waste stream may occur due to seasonal influences (beverage containers in resort areas, for example). In addition, changes in population, as experienced at seasonal resort areas, may have a marked consequence on the amounts and types of wastes generated.

4) Community recycling education programs.

5) Community mandated and enforced recycling programs versus voluntary recycling programs.

6) Relative ease (or difficulty) of participating in a curbside recycling program.

7) Tipping fee differentials between those at disposal facilities and at a MRF and/or the banning of some components (e.g., tires, yard

waste) from disposal facilities may bias the composition of the incoming waste stream at those facilities.

2.3.2 Mass Balance

2.3.2.1 Introduction and Preliminary Considerations

In addition to determining the relative quantities of the various components in the incoming waste streams in order to provide for the storage, separating, and processing of the raw materials as well as for the handling of residue and for the storage and shipping of finished products, it is necessary to determine the anticipated amounts of each of these components. The process by which this is accomplished is called a "Mass Balance" analysis. A proper Mass Balance analysis considers the nature of the incoming waste streams, the level of technology to be employed in the separation and processing of materials, the market specifications for the end products, the economic justification for separating and processing materials, and the legislated or project designated waste diversion levels which must be met. An important factor to consider before beginning the Mass Balance analysis and the subsequent sizing of the MRF and its sub-systems, is the anticipated total tonnage which the facility will be called upon to handle. It is imperative that this total be identified and defined as accurately as possible. Lacking such identification and definition, it is highly improbable that the facility will perform as designed.

Example: A waste characterization study and landfill records indicate that a community currently generates 46,500 tons per year of residential recyclables. Also, through the use of pilot programs and or knowledge of experiences in similar communities, it is estimated that 70% of the households may be expected to participate in a curbside collection program, then:

46,500	TPY residential recyclables generated
x 0.7	household participation rate
32,500	TPY maximum available for curbside collection

However, it is also realized that within each participating average household, only 80% of the possible recyclable materials will actually be placed at curbside for collection, then:

32,500	TPY maximum available
x 0.8	household internal participation rate
26,000	TPY collected at curbside and delivered to facility

A MRF may be planned to operate 52 weeks per year, 5 days per week, or 260 days per year. When a holiday falls on a weekday that day may be made up on the weekend, then:

$$\frac{26,000 \text{ TPY}}{260 \text{ days/year}} = 100 \text{ TPD collected at curbside and delivered to MRF.}$$

Care must be taken that anticipated growth in the population and the probable corresponding change in the waste stream quantities are allowed for. This does not mean that the Facility must be built to deal with the waste stream 20 years from now, but it does suggest that some planning needs to be carried out for that future requirement. Scheduled legislated waste diversion rates also have a bearing on Facility sizing.

If the Facility is to be properly sized, then the Facility capacity must be defined for the tipping floor as well as for the processing lines. In addition, the intended number of hours per day (e.g., 8,12,16) and days per week (e.g., 5,6,7) for receiving and processing waste must be defined.

From a purely economic standpoint, it is generally advantageous to utilize a Facility as continuously as possible at its design capacity. However, there are considerations which dictate that the Facility be operated less than 24 hours per day and/or 7 days per week, including:

- Traffic to and from the Facility
- Noise
- Time allocated for preventive maintenance
- Loss of efficiency on second and third shifts (If normal operation depends upon two or three shifts, there is less opportunity to make up for the inevitable down time.)
- Substantial changes in waste deliveries during the year (e.g., at resort areas).

2.3.2.2 Process Flow Chart and Mass Balance

For the purpose of example, it is assumed that the Facility will receive 100 TPD of paper and commingled recyclable containers in the proportions according to Table 2-1. The overall flows and mass balance may be represented on a Summary Flow Chart as illustrated in Figure 2-2. A 90% recovery rate has been assumed which results in 10 TPD of residue to be landfilled. The designer should realize that the greater the separation of material categories that occurs at the source, the higher the probable recovery rate of those recyclables at the MRF.

2.3.3 Technology Considerations for a Basic MRF

The separation and processing steps required or desired at a MRF, are influenced by market requirements, by the characteristics of the feedstock, and by the economics associated with separation, processing, and transportation. Additionally, in geographical areas where labor wages are historically low and unemployment high, there is greater reason to favor a labor intensive approach than there is in those areas where labor is scarce and labor wages high. The total quantity of materials and the relative percentages of material grades or categories will have an effect upon the methods employed for recovery and processing, and most certainly upon the equipment selected to recover and process the various materials.

Tables 2-2 and 2-3 address some of the more common design considerations of low and high technology systems. The combinations of low and high technologies are virtually without end.

2.3.4 Finished Product Specifications

A select list of grades and definitions adapted from the Scrap Specifications Circular 1990 as issued by the Institute of Scrap Recycling Industries Inc. (ISRI) is presented in Appendix 2. The material in Appendix 2 provides specifications for tin and aluminum cans. Finally, Appendix 2 presents examples of actual buyer specifications for various recyclable materials.

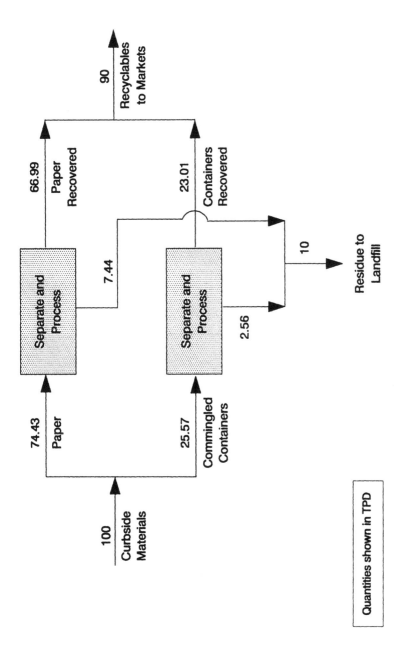

Figure 2-2. Summary flow chart and mass balance.

Table 2-2. Common Design Consideration of Low Technology Systems.

Material	Basic Feedstock	Tipping Floor	Infeed Conveyor	Sorting Conveyor (or room)	Interim Storage	Preparation for Shipping	Finished Product Storage
Paper	Mixed wet & dry paper, including newsprint, old corrugated containers (OCC), high-grades, books magazines, & contaminants	Handpick OCC & contaminants	Handpick OCC & contaminants	Handpick OCC, magazines, high-grades, mixed paper, etc.	In piles on processing floor or in bins	Ship loose, as is, or baled	In piles on processing floor, in bins, or compacted or baled in transport vehicles
Commingled Containers	Tin, bimetal, & aluminum cans, plastic, & glass containers, & contaminants	Handpick contaminants	Handpick contaminants; magnetic separator for ferrous	Handpick plastic, aluminum, contaminants	In piles, bins, or containers	Ship loose, as is	In piles, bins, containers, or transport vehicles

Table 2-3. Common Design Considerations of High Technology Systems.

Material	Basic Feedstock	Tipping Floor or Special Station	Infeed Conveyor	Trommel	Sorting Conveyor (or room)	Interim Storage	Preparation for Shipping	Finished Product Storage
Paper	Mixed wet & dry paper, including newsprint, old corrugated containers (OCC), high grades, books, magazines & contaminants	Handpick OCC & contaminants assisted by a grapple &/or front-end loader	Grapple or front-end loader	Separates oversize OCC &/or newspaper from mixed paper	Handpick remaining OCC, magazines, high-grade, & mixed paper, etc.	Accumulated in bins or bunkers before being selectively conveyed to baler	Auto-tie baler	In stacks or bales on processing floor or stacked in transport vehicle

Material	Basic Feedstock	Tipping or Special Station	Infeed Conveyor		Screen		Traveling Chain Curtain	
Commingled Containers	Tin, bimetal, & aluminum cans, plastic and glass containers, & contaminants	Handpick contaminants	Handpick contaminants; magnetic separator for ferrous		Broken glass as undersized materials		Separate aluminum and plastic from glass	

Table 2-3. Common Design Considerations of High Technology Systems. (CONTINUED)

Material	Sort	Bale	Biscuit	Shred	Air Classify	Store
Ferrous (Bimetal)	Manual separation of tin cans and bimetal (if required)	With auto-tie baler	With can densifier palletize			In stacks on processing floor outdoors, or in a transport vehicle
Ferrous (Tin Cans)	Manual separation of tin cans and bimetal (if required)	With auto-tie baler	With can densifier and palletize	With can shredder	To remove labels	Convey shredded cans to outside transport vehicle, or bales of biscuits in stacks of processing floor, outdoors, or in a transport vehicle.

Material	Separate	Flatten	Store	Bale	Store	Biscuit	Store
Aluminum	Eddy current apparatus separates aluminum from plastic	With can flattener	Pneumatically convey to outside transport vehicle	With auto-tie baler	In bales on processing floor or outdoors	Compress in a densifier and palletize	In stacks on the processing floor or outdoors or in a transport vehicle

Material	Sort	Interim Storage	Perforate	Bale	Store
Plastic (PET)	Manual sort of PET, HDPE, other	In overhead hoppers	Drop from overhead hopper or pneumatically convey to perforator	Mechanically or pneumatically from perforator to auto-tie baler	In stacks or bales on processing floor or outdoors in transport vehicles

Table 2-3. Common Design Considerations of High Technology Systems. (CONTINUED)

Material	Sort	Interim Storage	Granulate	Bale	Store
Plastic (HDPE)	Manual sort of PET, HDPE, other	In overhead hoppers	Drop from overhead hopper or pneumatically convey to granulator	Mechanically or pneumatically convey from hopper to auto-tie baler	Granulated in gaylords on processing floor before loading into transport vehicle, baled in stacks on processing floor or outdoors in transport vehicles.

Material	Sort	Crush	Upgrade	Store
Glass	Optical automatic sort or hand sort by color	To meet market specifications	Remove paper labels, metal lids, & other contaminants by trommel and/or air classifier	In bunkers for loading by front-end loader, or in overhead bins for selectively conveying to transport vehicles

2.3.5 Flow Chart and Mass Balance -- Low Technology

Using the proportions of recyclable materials as shown in Table 2-1, the Flow Chart and Mass Balance for the Paper Line are shown in Figure 2-3. Similarly, Figure 2-4 is a Flow Chart and Mass Balance for the Commingled Container Line. With reference to either of these two figures the reader will note that more or fewer separations may take place at the sorting station. In the case of the Paper Line, Figure 2-3, it is possible that the only product desired is "mixed paper" in which case it is only necessary to remove whatever material is considered to be contaminating to the product. Or, as markets develop or change, it may prove to be of value to separate corrugated, office paper, or mixed paper products. The system can be designed to accommodate such changes with minimal capital expenditure.

In the case of the Commingled Container Line, Figure 2-4, it is also possible to increase or decrease the number of materials categories separated from the incoming stream. For example, glass can be sorted by color if warranted, or tin cans, if no market is available, are permitted to join the residue to be landfilled. In each case depicted in Figures 2-2 through 2-4, a 90% material recovery rate has been assumed. In actual practice there are many factors which have an influence on this recovery rate. Some of these factors are listed in Table 2-4.

2.3.6 Flow Chart and Mass Balance -- High Technology -- Paper Line

Using the proportions of recyclable materials as shown in Table 2-1, the Flow Chart and Mass Balance for the Paper Line are shown in Figure 2-5. With reference to this Flow Chart, the incoming paper is conveyed to a trommel (a rotating cylindrical screening device) so designed that large material (newspaper and corrugated) will pass through the cylinder (oversize material, or "overs") while small material (mixed paper) will fall through the screen openings as undersized material, ("unders"). Provision should be made at the trommel inlet to manually divert extra large pieces of corrugated which may jam the system. It should be noted that the trommel may also be designed to remove grit and gravel and other components smaller than mixed paper that would contaminate the end product.

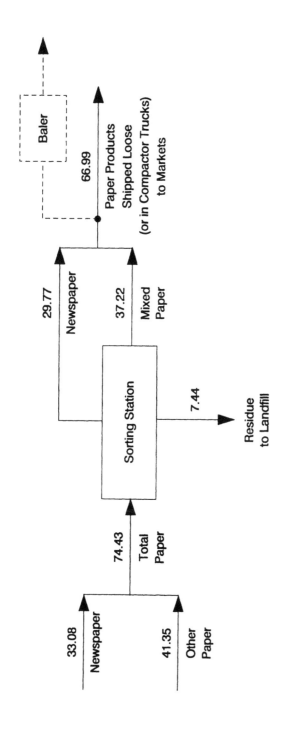

Figure 2-3. Flow chart and mass balance--low technology--paper line.

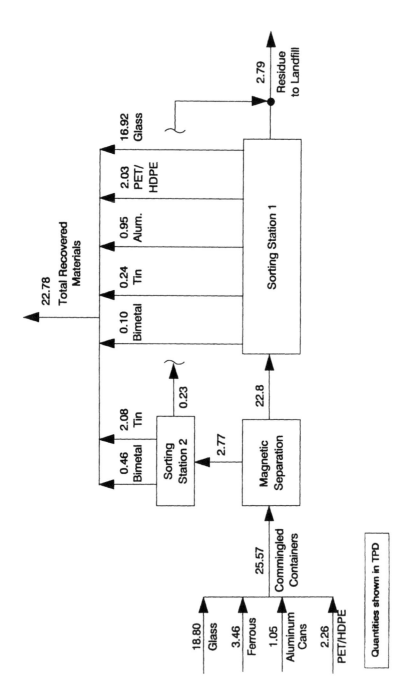

Figure 2-4. Flow chart and mass balance--low technology--commingled container line.

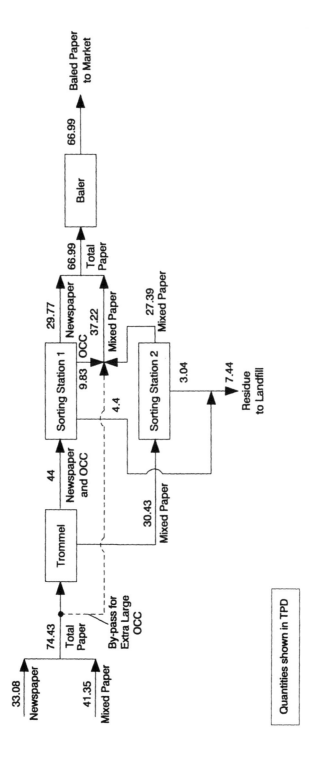

Figure 2-5. Flow chart and mass balance--high technology--paper line.

Table 2-4. Factors Affecting Material Recovery Rate

Factor	Explanation
Market Specifications	"Loose" (i.e., incomplete) specifications potentially increase recovery rates over those recovery rates that are attainable in the case of "tight" (i.e., complete) specifications
Contamination of Incoming Materials	This factor is closely related to that of market specifications in that some markets will accept products which other markets consider as unacceptable because of contamination
Glass Breakage	This factor applies to glass containers and it is influenced by the manner in which the containers are set out, collected, transported, sorted, and handled at the Facility. Broken glass is more difficult to sort than unbroken glass
Relative Quantities per Sorter	Over a given period of time the greater the number of units of any given recyclable a sorter must separate from the waste stream, the lower the recovery rate. Conversely, recovery can be increased by increasing the number of sorters utilized
Equipment Design	Proper design of conveyors and separation equipment for the types and quantities of materials handled, directly affects recovery rates. For example, an excessive bed depth of commingled containers on a conveyor can substantially limit the manual or automatic recovery of any given material
Human Factors	Providing a clean, well-lit, and pleasant environment in which to work with particular attention to worker training, safety, health, and comfort will tend to increase recovery rates
Fictitious Weights	Most MRF's are now equipped with a truck scale and scale house. Additionally, values for tare weights for regular haul vehicles are either in the scale's computer data base or they are determined after tipping. Care must be taken, particularly during periods of inclement weather, that the weight of an incoming load does not include an inordinate amount of free water from a recent rain or snowstorm which would inaccurately represent the weights of incoming materials.

At Sorting Station #1, unacceptable materials are removed for landfilling. In addition, corrugated and mixed paper are separated from the newspaper stream. The trommel "unders" (i.e., mixed paper) are conveyed to Sorting Station #2 where unacceptable materials are removed for landfilling. The remaining paper joins the corrugated removed from

Sorting Station #1. Newspaper and mixed paper are collected separately and accumulated in individual bins and then conveyed to an auto-tie baler for shipment to markets.

As a variant to this scheme, the trommel as well as Sorting Station #2 could be eliminated. This option, of course, would put a greater burden on the manual separation effort at Sorting Station #1.

2.3.7 Flow Chart and Mass Balance--High Technology--Commingled Container Line

Continuing with the example, and using the proportions of recyclable materials as shown in Table 2-1, the Flow Chart and Mass Balance for the Commingled Container Line are shown in Figure 2-6.
With reference to this Flow Chart, the incoming commingled containers are introduced to the processing line on a common infeed conveyor. For the purposes of describing the processes as well as to provide a modular system concept, the commingled container line is presented as consisting of four basic modules. They are:

- Ferrous module
- Glass module
- Plastics module, and
- Aluminum module

2.3.7.1 Flow Chart -- High Technology -- Ferrous Module

Flow Chart, Figure 2-7, is an enlarged view of that portion of Figure 2-6 which pertains to the separation and processing of ferrous material. All commingled containers are conveyed to a magnetic separator whose function is to extract all ferrous materials from the rest of the commingled container stream. The efficiency with which this task is accomplished is a function of the design of the magnetic separator, the bed depth of the materials subjected to the magnetic field, the ratio of ferrous containers to other materials and the proportion of ferrous containers which are filled or partially filled with food, liquid or other substances.

Once separated from the other containers and depending upon the markets, the ferrous containers are manually sorted (Sorting Station #1) into two streams, i.e., bimetal and tin. Residue is collected and transported to landfill. Bimetal containers may be flattened, baled, or

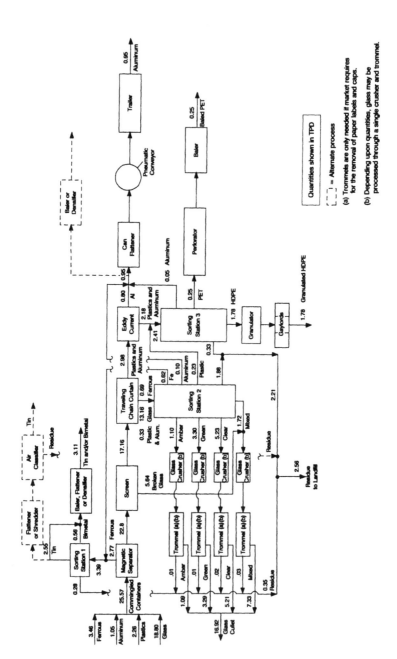

Figure 2-6. Flow chart and mass balance--high technology--commingled container line.

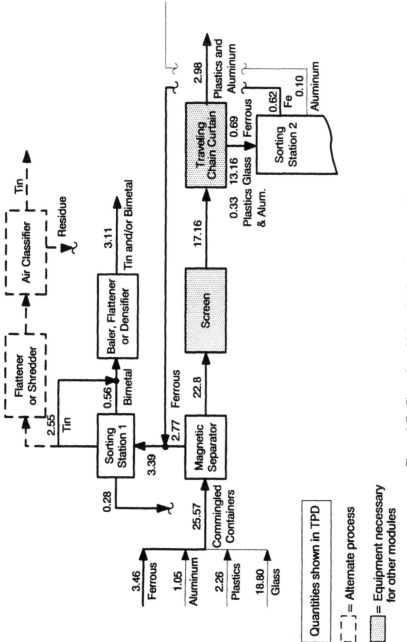

Figure 2-7. Flow chart--high technology--ferrous module.

densified into biscuit form. Tin cans may be flattened or shredded and introduced to an air classifier for the removal of labels loosened by the flattening, or shredding process. Alternatively, tin cans may be flattened, baled or densified with or without bimetal cans. Ferrous cans that are not removed by the magnetic separator from the commingled containers stream are conveyed to a sorting station where manual separation takes place. The cans removed manually are returned, by means of conveyors, to join the ferrous removed by the magnetic separator.

2.3.7.2   Flow Chart--High Technology--Glass Module

Flow Chart, Figure 2-8, is an enlarged view of that portion of Figure 2-6 which pertains to the separation and processing of glass. After magnetic separation of ferrous from the commingled container stream, the remaining containers pass over a screen which enables much of the broken glass to be removed as "unders." The "overs" enter a traveling chain curtain which separates plastic and aluminum containers from the glass containers. The glass containers are then conveyed to a sorting station. Glass containers are sorted by color with each color passing through a glass crusher. Depending upon market specifications each cullet stream may be introduced to a small trommel for removal of paper labels and caps. The mechanical removal of labels and caps may be further assisted by pneumatic means. The "unders" from the screening operation join the mixed glass from the sorting station and are processed in the same manner as are the various colored glass containers.
Residues from the sorting station and the trommels are collected and transported to landfill.

2.3.7.3 Flow Chart -- High Technology -- Plastics Module

Flow Chart, Figure 2-9, is an enlarged view of that portion of Figure 2-6 which pertains to the separation and processing of plastics. After magnetic separation of ferrous and the removal of broken glass by the screen, plastic and aluminum containers are separated from the glass containers by means of a traveling chain curtain. An eddy current device is then used to eject aluminum cans from the plastic/aluminum substream. The plastic containers are conveyed to a sorting station where PET is separated from HDPE. Trace plastics entrained with the glass substream from the air classifier or traveling chain curtain are separated at the glass sorting station and transferred to the plastics sorting station for PET/HDPE separation.

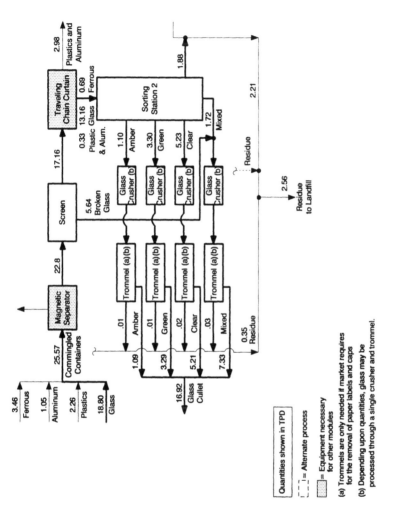

Figure 2-8. Flow chart--high technology--glass module.

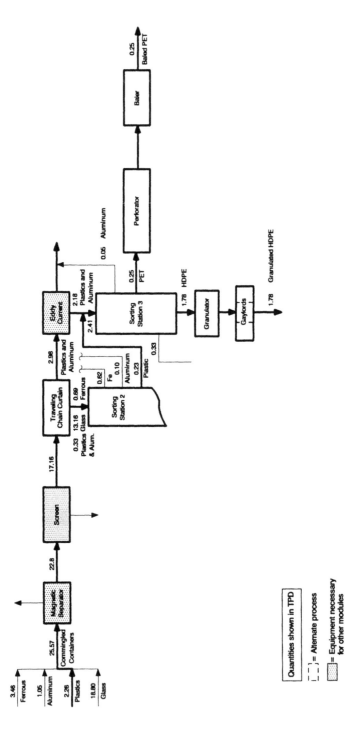

Figure 2-9. Flow chart--high technology--plastics module.

PET containers are collected, perforated and baled. HDPE containers are granulated. The plastic granules are loaded into gaylords for shipment to market. Residue is collected and transported to landfill.

2.3.7.4 Flow Chart -- High Technology -- Aluminum Module

Flow Chart, Figure 2-10, is an enlarged view of that portion of Figure 2-6 which pertains to the separation and processing of aluminum cans. After separation of aluminum cans from plastic containers by the eddy current device which employs an electromagnetic field to repel non-ferrous metals, the cans are flattened and pneumatically conveyed to a transport trailer. Alternatively, the cans may be baled or densified into biscuit form to meet market specifications. Trace aluminum which may have escaped separation from the plastic by the eddy current device is routed from the plastics sorting station to the can flattener, baler or densifier as applicable.

2.3.8 Flow Charts/General Comment

With regard to the flow charts illustrated in Figures 2-3 through 2-10 for low and high technology systems, the reader should recognize that there are almost limitless combinations and modifications of the systems presented. For example, Figure 2-6 includes a traveling chain curtain (or other automatic sorting device) to sort glass from the rest of the waste stream. If this operation did not exist, then the screen "overs" would be directed to the eddy current device for aluminum extraction with the remainder directed to a sorting station which would combine the activities described as taking place at Sorting Stations #2 and #3.

2.3.9 Material Densities

In order to properly size a MRF and to select or design the equipment used therein, it is necessary to have knowledge of the densities associated with the various materials as received, handled, processed, and stored. All density values are the result of dividing the weight of the material by its volumes. The differences arise due to the forms in which the material is found.

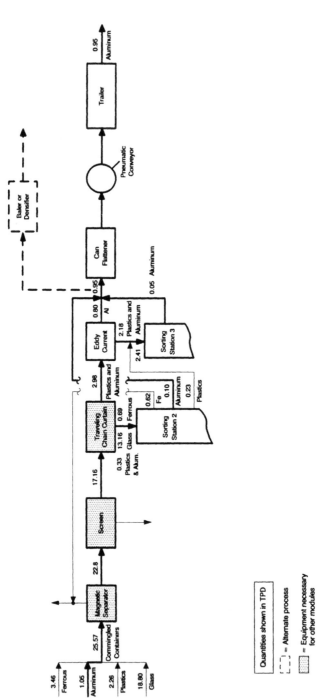

Figure 2-10. Flow chart--high technology--aluminum module

MATERIAL DENSITY DEFINITIONS:

- Bulk Density: Weight of material divided by the volume of that portion of a container which is filled with the material
- True Density: Weight of the material in its natural form (e.g., glass, rather than glass bottles) divided by its volume.
- Compressed Density: Weight of material divided by its volume during or after having been exposed to compressive forces in a confined space. Cellulosic materials can be compressed to densities as high as 75 lb/cu ft.

Densities of several materials received, handled, processed, and stored at MRFs are listed in Table 2-5.

2.3.10 Fixed Equipment

The purpose of this sub-section is to provide guidance to the reader who is involved in the review and selection process of fixed equipment as employed in a MRF.

2.3.10.1 Fixed Equipment Commonly Present in a MRF

A comprehensive list of various types of fixed equipment which may be included in a MRF is presented in Table 2-6.

2.3.10.2 Fixed Equipment Descriptions

The following equipment descriptions are provided to give the reader a brief overview of machinery commonly employed in a MRF. Since new special purpose machines continue to be developed to serve this growing industry, the list should not be regarded as all-inclusive. The Facility planner/designer should be particularly cautious in placing reliance upon unproven technology.

In the review and selection process of individual items of fixed equipment, it should be recognized that these items must not only compatibly interrelate with one another but also with the various collection vehicles which deliver the incoming materials as well as in-plant rolling equipment and transport vehicles for shipping the final products.

Table 2-5. Average Densities of Refuse Components

Component	Density
Refuse Densities	lb/yd^3
Loose	100-200
After dumping from compactor truck	350-400
In compactor truck	500-700
In landfill	500-900
Shredded	600-900
Baled in paper baler	800-1200
Bulk Densities	lb/ft^3
OCC	1.87
Aluminum cans	2.36
Plastic containers	2.37
Miscellaneous paper	3.81
Garden waste	4.45
Newspaper	6.19
Rubber	14.90
Glass bottles	18.45
Food	23.04
Tin cans	4.90
True Densities	lb/ft^3
Wood	37
Cardboard	43
Paper	44-72
Glass	156
Aluminum	168
Steel	480
Polypropylene	56
Polyethylene	59
Polystyrene	65
ABS	64
Acrylic	74
Polyvinylchloride (PVC)	78
Resource Recovery Plant Products	lb/ft^3
dRDF	39
Aluminum scrap	15
Ferrous scrap	25
Crushed glass	85
Powdered RDF (Eco-Fuel)	27
Flattened aluminum cans	9
Flattened ferrous cans	31

Table 2-6. Fixed Equipment Which May be Employed in a Materials Recovery Facility.

Material Handling Equipment
Belt Conveyor
Screw Conveyor
Apron Conveyor
Bucket Elevator
Drag Conveyor
Pneumatic Conveyor
Vibrating Conveyor

Separating Equipment
Magnetic Separator
Eddy Current Device (aluminum separator)
Disc Screen
Trommel Screen
Vibrating Screen
Oscillating Screen
Traveling Chain Curtain
Air Classifier

Size Reduction Equipment
Can Shredder
Can Densifier/Biscuiter
Can Flattener
Glass Crusher
Plastics Granulator
Plastics Perforator
Baler

Environmental Equipment
Dust Collection System
Noise Suppression Devices
Odor Control System
Heating, Ventilating, & Air Conditioning (HVAC)

Other Equipment
Fixed Storage Bin
Floor Scale for Pallet or Bin Loads
Truck Scale
Belt Scale

2.3.10.2.1 Material handling equipment (conveyors). The most common piece of equipment for handling materials in a MRF is the conveyor. There are several types of conveyors available. Selection of the correct types of conveyors in a MRF must take into consideration a number of interrelated factors. Complete engineering data are available for many types of conveyors. Consequently, their performance can be accurately predicted when they are used for handling materials having well-known characteristics. However, if the material characteristics are not well-known, even the best designed conveyor would not perform well. Some of the most important factors to be considered in conveyor selection include:

- Capacity
- Length of travel
- Lift
- Characteristics of the material
- Cost

The most common types of conveyors used in a MRF are the belt conveyor, the apron conveyor, and the screw conveyor. A short description of each follows.

Belt Conveyor

In a MRF, the belt conveyor is employed in several forms. Some of these forms include:

Trough Type: In general, the trough type belt conveyor will use troughing idler rolls which cause the conveyor belt to form a concave contour with its sides sloping at 20°, 35°, or 45° with a horizontal plane (see Figure 2-11). The purpose of this cross-sectional concavity is to retain free flowing materials (e.g., aluminum cans, bottles, crushed glass, etc.) and to minimize or prevent spillage. In order to further minimize spillage problems, skirt boards (see Figure 2-12) are often used at belt transfer points.

The Conveyor Equipment Manufacturers Association (CEMA) provides a design handbook for belt conveyors. Tables 2-7 and 2-8 have been adapted from information published by the CEMA for some specific materials generally handled in a MRF.

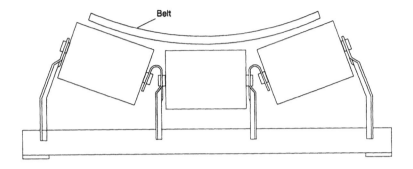

Figure 2-11. Trough type belt conveyor

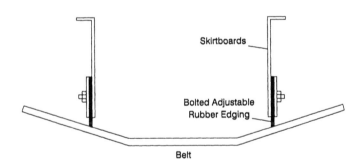

Figure 2-12. Belt conveyor with skirtboards

Table 2-7. Approximate Conveyor Belt Capacities (20° Trough)[1,6] (TPH)

Component[3]	Belt Width (Inches)[2]					
	18	24	36	48	60	72
Glass Bottles[4]	6.0	11.6	28.3	52.2	83.4	121.8
Plastic Bottles[4]	0.8	1.5	3.7	6.8	10.8	15.8
Aluminum Cans[4]	0.8	1.5	3.7	6.8	10.8	15.8
News[5]	3.9	7.5	18.1	33.3	53.1	77.5
OCC[5]	1.2	2.2	5.3	9.8	15.6	22.8
Loose Refuse[5]	3.5	6.7	16.0	29.4	46.9	68.4
Refuse from Compactor Truck[5]	8.7	16.6	40.0	73.6	117.2	171.0

EXAMPLE: To find capacity at other belt speeds: New belt speed = 20 FPM; Plastic Bottles, 36 in. belt width; TPH = 20 FPM/100 FPM x 3.7 TPH = 0.7 TPH

---

[1]Conveyor Speed = 100 FPM
[2]Edge Distance (inches) = 0.055 x belt width + 0.9. Three equal idler roll lengths
[3]Densities as per Table 2-5
[4]Surcharge Angle = 5°
[5]Surcharge Angle = 30°
[6]Based on capacities published in CEMA "Belt Conveyors for Bulk Materials"

Table 2-8. Approximate Conveyor Belt Capacities (35° Trough)[1,6] (TPH)

Component[3]	Belt Width (Inches)[2]					
	18	24	36	48	60	72
Glass Bottles[4]	8.9	17.2	41.7	77.0	122.9	179.4
Plastic Bottles[4]	1.2	2.2	5.4	10.0	16.0	23.3
Aluminum Cans[4]	1.2	2.2	5.4	10.0	16.0	23.3
News[5]	4.7	9.0	21.6	39.7	66.3	92.2
OCC[5]	1.4	2.6	6.3	11.7	19.5	27.1
Loose Refuse[5]	4.1	7.9	19.0	35.0	58.5	81.3
Refuse from Compactor Truck[5]	10.4	19.8	47.6	87.5	146.2	203.3

EXAMPLE: To find capacity at other belt speeds: New belt speed = 20 FPM; Plastic Bottles, 36 in. belt width; TPH = 20 FPM/100 FPM x 5.4 TPH = 1.1 TPH

---

[1]Conveyor Speed = 100 FPM
[2]Edge Distance (inches) = 0.055 x belt width + 0.9. Three equal idler roll lengths
[3]Densities as per Table 2-5
[4]Surcharge Angle = 5°
[5]Surcharge Angle = 30°
[6]Based on capacities published in CEMA "Belt Conveyors for Bulk Materials"

The designer is referred to the most recent issue of ASME/ANSI B20-1, Safety Standard for Conveyors and Related Equipment, for information and guidance in the design construction, installation, operation, and maintenance of conveyors and related equipment. In addition to general safety standards applicable to all conveyors and related equipment, Section 6.1 of the Standard is specifically applicable to belt conveyors.

Flat Belt Type: Most flat belt conveyors employed in a MRF are of the "slider belt" design in which the conveyor belt is backed up by and slides on a steel supporting surface rather than on idler rolls.

Flat belt conveyors are popularly utilized in the sorting process at a MRF for they permit easy access to the material carried on the belt.

When a flat belt conveyor is used in an inclined position, it is often supplied with cleats and skirt boards for the full length of the conveyor in order to more positively convey the materials and prevent spillage.

Tables 2-9 and 2-10 have been adapted from belt capacity tables published by the CEMA for some specific materials generally handled in a MRF.

Most flat belt conveyors rely upon the friction force between the head pulley (drive pulley) and the conveyor belt to drive the conveyor. Where particularly heavy loads are anticipated (e.g., MSW), chains are attached to the underside and to each side of the belt for the full length. This configuration usually is accompanied by cleats attached to the carrying surface of the belt as well as full length skirtboards to retain material on the conveyor. ASME/ANSI B 20.1 Safety Standard is equally applicable to flat belt conveyors as it is to the trough type as previously discussed.

Apron Conveyor

An apron conveyor consists of steel pans (flat or contoured) supported by chains and is used in applications in which the conveyor may be subject to substantial impact and abuse. Guide rollers riding on steel rails minimize the frictional forces. Cleats may be incorporated on the pans for inclined applications. Apron conveyors are often employed as infeed conveyors and may be located in a pit below floor level. Ample provision should be made for access for cleanout and maintenance. Section 6.5 of the ASME/ANSI B 20.1 Safety Standard is specifically applicable to apron conveyors.

Table 2-9. Approximate Conveyor Belt Capacities (Flat Belt)[1,6] (TPH)

Component[3]	Belt Width (Inches)[2]					
	18	24	36	48	60	72
Glass Bottles[4]	1.1	2.2	5.1	9.4	14.9	21.8
Plastic Bottles[4]	0.1	0.3	0.7	1.2	1.9	2.8
Aluminum Cans[4]	0.1	0.3	0.7	1.2	1.9	2.8
News[5]	2.4	4.6	10.9	19.9	31.6	46.1
OCC[5]	0.7	1.3	3.2	5.9	9.3	13.6
Loose Refuse[5]	2.1	4.0	9.6	17.6	27.9	40.7
Refuse from Compactor Truck[5]	5.3	10.0	24.0	43.9	69.8	101.7

EXAMPLE: To find capacity at other belt speeds: New belt speed = 20 FPM; Plastic Bottles, 36 in. belt width; TPH = 20 FPM/100 FPM x 0.7 TPH = 0.14 TPH

---

[1]Conveyor Speed = 100 FPM
[2]No idlers
[3]Densities as per Table 2-5
[4]Surcharge Angle = 5°
[5]Surcharge Angle = 30°
[6]Based on capacities published in CEMA "Belt Conveyors for Bulk Materials"

Table 2-10. Approximate Conveyor Belt Capacities (Flat Belt with 6-in. High Skirtboards)[1,6] (TPH)

Component[3]	Belt Width (Inches)[2]					
	18	24	36	48	60	72
Glass Bottles[4]	34.0	47.4	75.1	104.1	134.4	165.9
Plastic Bottles[4]	4.4	6.2	9.8	13.5	17.5	21.6
Aluminum Cans[4]	4.4	6.2	9.8	13.5	17.5	21.6
News[5]	13.6	19.9	34.7	52.1	72.3	95.1
OCC[5]	4.0	5.9	10.2	15.3	21.3	28.0
Loose Refuse[5]	12.0	17.6	30.6	46.0	63.8	83.9
Refuse from Compactor Truck[5]	30.0	44.0	76.5	114.9	159.4	209.8

EXAMPLE: To find capacity at other belt speeds: New belt speed = 20 FPM; Plastic Bottles, 36 in. belt width; TPH = 20 FPM/100 FPM x 9.8 TPH = 2.0 TPH

[1]Conveyor Speed = 100 FPM
[2]No idlers
[3]Densities as per Table 2-5
[4]Surcharge Angle = 5°
[5]Surcharge Angle = 30°
[6]Based on capacities published in CEMA "Belt Conveyors for Bulk Materials"

Screw Conveyor

The screw conveyor (or auger) may be used to transport dry, dense, free flowing materials (e.g., tin cans formed as nuggets). Screw conveyors have also been used for bin discharge and as metering feed devices. These units are not designed to transport stringy, abrasive, or very wet materials.

Pneumatic Conveyor

A pneumatic conveyor utilizes a stream of air to convey suspendable materials (e.g., aluminum cans or dust) through a tube. Pneumatic conveyors may utilize either a vacuum or a positive pressure. The pneumatic conveyor offers the facility designer more flexibility in equipment location. However, the number of changes in direction in the lines should be kept to a minimum since they result in pressure (efficiency) losses as well as probable points of stoppages and wear.

Separation Equipment

The following equipment is employed to separate one or more materials from the waste stream or sub-stream. It should be recognized that none of these devices can be expected to be 100% effective.

Magnetic Separator

Magnetic separation is a relatively simple unit process and is used to recover ferrous metal from the commingled waste stream. Magnets may be either of the permanent or the electromagnetic type. They are available in three configurations, namely, the drum (Figure 2-13), the magnetic head pulley (Figure 2-14), and the magnetic belt (Figure 2-15). They may be assembled and suspended in line, crossbelt, or mounted as conveyor head pulleys. The magnetic head pulley conveyor is arranged so that in its operation, the material to be sorted is passed over the pulley in such a manner that the nonferrous material will fall along a different trajectory than will the ferrous material. The drum magnet assembly can be installed for either overfeed or underfeed and directs the ferrous along a trajectory other than that taken by the nonferrous material. The magnetic belt, in its simplest form, consists of single magnets mounted between two pulleys that support a cleated conveyor belt mechanism. The efficiency of magnetic separation is affected by the bed depth of the

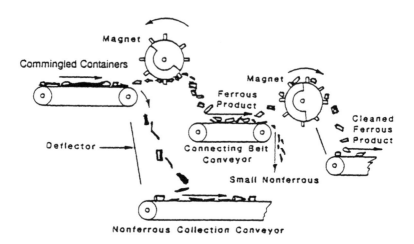

Figure 2-13. Multiple magnetic drum.

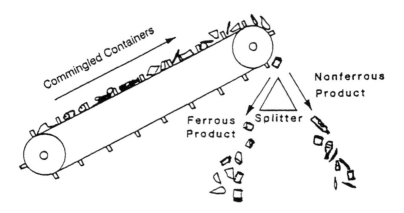

Figure 2-14. Magnetic head pulley

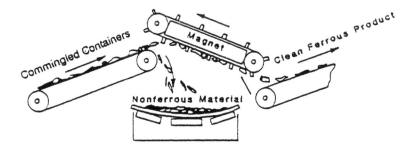

Figure 2-15. Magnetic belt.

waste stream. For more complete removal of ferrous, a secondary magnetic separator may be considered. Conveyor and hopper components in the vicinity of the magnetic field should be constructed of non-magnetic materials. Additional information on magnetic separation can be obtained in References 1 to 7.

Eddy Current Device (Aluminum Separator)

An aluminum separator employs either a permanent magnetic or electromagnetic field to generate an electrical current (eddy) which causes aluminum cans (non-ferrous metals) to be ejected and separated from other materials. Aluminum separation may take place in the form of a conveyor head pulley or in the form of an inclined stainless steel plate. Additional information on aluminum separation can be found in References 7 to 10.

Disc Screen

A disc screen consists of parallel multiple shafts all rotating in the same direction. Discs are mounted on each of these shafts, and spaced in such a fashion so that the discs on one shaft are located midway between the discs on an opposing shaft. The shafts and discs are so positioned relative to each other as to establish fixed interstices through which the undersize material (e.g., broken glass or grit) will pass and the oversize material is conveyed by both the discs and the series of rotating shafts. A schematic view of a disc screen is presented in Figure 2-16.

Disc screens are subject to damp and stringy material wrapping around the shafts and discs and thus reducing the interstices. At the infeed location, abrasive material (e.g., broken glass or grit) may abrade the outside diameters of the shafts and discs so as to substantially increase the interstices. Also, large pieces of corrugated may act as a barrier to smaller material dropping through the interstices.   Any of these conditions can have a significant detrimental effect upon performance.

Trommel Screen

The trommel is a rotary cylindrical screen, generally downwardly inclined, whose screening surface consists of wire mesh or perforated plate. A diagram of a typical trommel screen is presented in Figure 2-17. The tumbling action of the trommel efficiently brings about a separation of individual items or pieces of material that may be attached to each other, or even of one material contained within another. Large trommels (8 to 10 ft in diameter and up to 50 ft long) have been used to separate large OCC and/or newsprint from mixed paper or commingled containers (particularly from glass containers). Small trommels (1 to 2 ft in diameter by 2 to 4 ft long) have been used to separate labels and caps from crushed glass. These small units are sometimes used in conjunction with an air stream to aid in the separation.

Two stage trommels have also been used in waste processing. In two stage trommels, the first stage (the initial length of screen) is provided with small apertures (e.g., 1 in. diameter) which permit broken glass, grit, and other small contaminants to be removed. The second stage is provided with larger apertures (e.g., 5 in. diameter) which allow glass, aluminum, and plastic containers to be removed from the waste stream. In the particular types of MRFs discussed in this document, the oversize materials (overs) might consist primarily of OCC and news, depending upon the make-up of the incoming waste stream.

Many factors influence the separation efficiency of a trommel including:

- Characteristics and quantity of the incoming materials
- Size, proportions, and inclination of the cylinder screen
- Rotational speed
- Size and number of screen openings

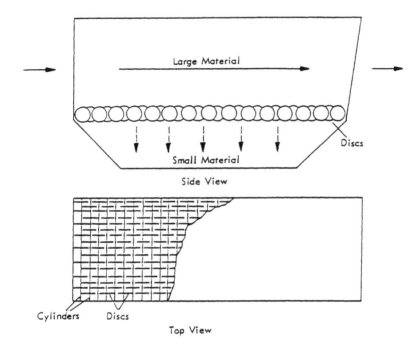

Figure 2-16. Disc screen

Vibrating Screen

A vibrating screen utilizes a wire mesh or perforated plate screen deck to separate relatively dense, dry, undersize materials from less dense oversize materials. A schematic diagram of a vibrating screen is given in Figure 2-18. Vibrating conveyors are more tolerant of stringy materials than are other conveyors.

Damp sticky materials have a tendency to blind the screen deck and thus impair the performance. Large pieces of corrugated and/or excessive material bed depth can substantially decrease separation efficiency.

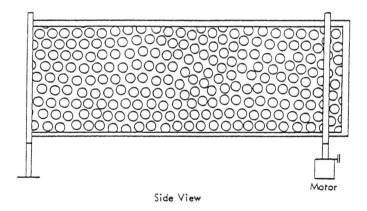

Side View

Motor

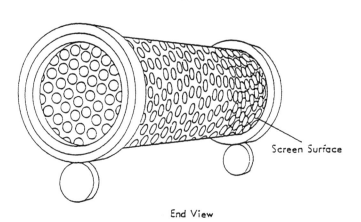

Screen Surface

End View

Figure 2-17. Trommel screen.

Oscillating Screen

An oscillating screen is configured in a similar fashion as a vibrating screen except that the motion is of an orbital nature in the plane of the screen deck. The same comments as those presented in section for vibrating screens apply.

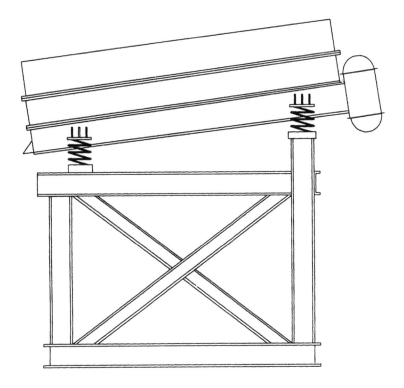

Figure 2-18. Vibrating screen.

Traveling Chain Curtain

The traveling chain curtain consists of one or more rows of common chain each suspended from a continuously revolving link-type conveyor chain describing a somewhat elliptical orbit around a vertical axis. The curtain provides a barrier to less dense (e.g., aluminum and plastic) containers while permitting denser material (e.g., glass) to pass through on a downwardly inclined surface. The efficiency of the traveling chain curtain can be greatly influenced by the feed rate into the unit. Excessive quantities of incoming material may cause lighter materials to push through the curtain rather than to be directed to one side. Detailed discussions about screens commonly used in the waste processing field can be found in References 6, 7, and 11 to 16.

Air Classifier

Air classification employs an air stream to separate a light fraction (e.g., paper and plastic) from a heavy fraction (e.g., metals and glass) in a waste stream. Variables other than density such as particle size, surface area, and drag also affect the process of material separation through air classification. Consequently, aluminum cans, by virtue of a high drag-to-weight ratio, may appear in the light fraction and wet and matted paper may appear in the heavy fraction.

Air classifiers may be provided in one of a number of designs. The vertical, straight type is one of the most common units. Air classifiers require provisions for appurtenant dust collection, blower, separation, and conveying. Schematic diagrams of typical air classifiers are provided in Figure 2-19.

A considerable amount of work has been carried out in the area of air classification of solid wastes. Results of some of this work are reported in References 7 and 17 to 22.

2.3.10.2.3   Material handling equipment (size reduction).
Several types of size reduction equipment are used for waste processing. The equipment is employed to reduce the particle size and/or increase the bulk density of material in order to meet market specifications and/or to reduce the cost of storage and transportation.

Can Shredder

A can shredder is employed to reduce aluminum cans to particles of small size (less than 1 in.). The process increases the density and thereby conserves on transportation costs. Shredded aluminum may command a premium price. The shredder is often supplied complete with infeed conveyor, magnetic separator, blower, and dust collector. Due to the costs involved in size reduction, prior to the installation and operation of a can shredder it is especially important to determine if the specifications call for shredded aluminum.

Can Densifier/Biscuiter

A can densifier is used to form aluminum cans into biscuits generally weighing on the order of 40 lb each. The capacity of a densifier may be increased by placing the densifier in series with and following a can

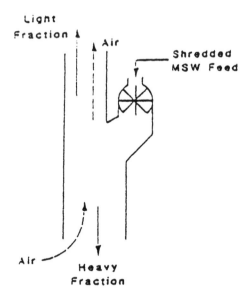

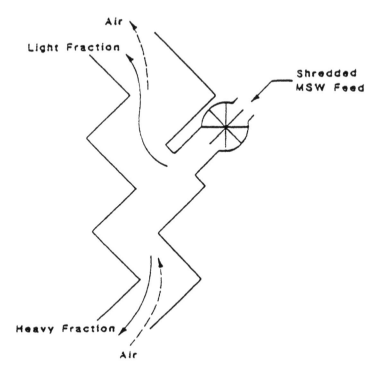

Figure 2-19. Vertical air classifiers.

flattener. A densifier offers a viable option to baling aluminum cans. As with the can shredder, it is important to verify that the market will accept and pay for the biscuit-shaped product. The typical range of dimensions for aluminum can densifiers commonly used in MRFs is illustrated in Figure 2-20. Production rates as a function of horsepower for aluminum can densifiers are presented in Table 2-11.

Can Flattener

A can flattener is a device used for flattening aluminum or tin cans. It is often provided complete with inlet hopper, belt conveyor, magnetic separator and pneumatic discharge. The crushing mechanism generally consists of a steel drum with hardened cleats rotating against a pressure plate or vulcanized rubber pressure drum or one or more sets of steel crushing rolls or drums. Overload protection and provisions for separating any liquids that may still be in the containers should be incorporated in the system design.

Figure 2-21 illustrates the typical range of dimensions for can flatteners (with infeed conveyors) as commonly used in MRFs. Clearance should be provided for maintenance although most flatteners are relatively light and portable and thus they can readily be moved to another location for maintenance if necessary. Typical production rates as a function of horsepower for aluminum and steel can flatteners are presented in Table 2-12.

Glass Crusher

A glass crusher is used to reduce whole glass bottles to small particle sizes in order to meet market specifications. Glass crushers are often supplied with a feed hopper and conveyor. Glass crushers are units that typically require relatively high maintenance because of the abrasive nature of the glass. Specifications from the users should be checked before glass crushers are included in the design of a MRF since some buyers prefer to perform their own crushing. Glass crushing is a dust producing operation and provision should be made to address this condition.

Figure 2-22 illustrates the typical range of dimensions for glass crushers used in MRFs. Clearance should be provided for maintenance although most crushers are relatively light and portable and thus could be

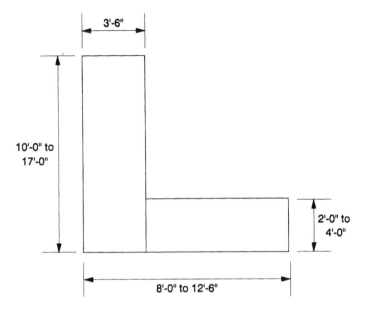

Figure 2-20. Typical range of dimensions for can densifiers.

Table 2-11. Typical Production Rates (lb/hr) and Horsepower for Aluminum Can Densifiers.

Lb/hr	Wt. of Biscuit (lb)	HP
300-500	18	5
600-900	18	7-1/2
2500 unflattened 3600 flattened	40	25

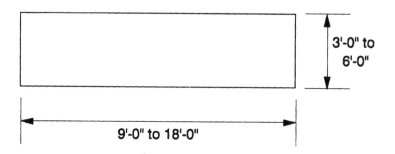

Figure 2-21. Typical range of dimensions for can flatteners.

Table 2-12. Typical Production Rates (lb/hr) and Horsepower for Aluminum Can Flatteners

	Horsepower		
Lb/hr	Blower	Flattener	Conveyor
Aluminum			
1,200 unflattened	5	5	1/3
2,000 unflattened 4,000 flattened	5	7.5	1/2
Steel			
2,000 unflattened		7.5 to 10	1/2

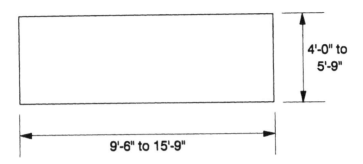

Figure 2-22. Typical range of dimensions for glass crushers (with infeed conveyor).

readily moved, if necessary, to another location for maintenance. Typical production rates versus horsepower for glass crushers commonly used in MRFs are presented in Table 2-13.

Plastics Granulator

A plastics granulator is used to size reduce PET and/or HDPE containers to a flake-like condition. The granulated plastic is generally shipped in gaylords. Due to the relatively large reduction in volume, substantial savings in shipping can be realized when plastic granulation is employed. Plastics granulation is an operation that requires a relatively high degree of maintenance and may be prone to dust generation. As with crushed glass, markets should be checked to verify that the specifications call for granulated material. Some potential buyers may wish to maintain close control over the type of plastic they receive and believe that they are better able to do so by requiring that the plastic be baled rather than granulated.

Plastics Perforator

Technically, a plastics perforator is not classified as a piece of size reduction equipment. However, its use is so intimately associated with that of a baler that it is included in this discussion. A plastics perforator

Table 2-13. Typical Production Rates (TPH) and Horsepower for Glass Crushers.

TPH[a]	Horsepower	
	Crusher	Conveyor
1-3	1	1/3
3-4	1 to 2	1/2
5-6	1 to 2	1/2
15	7-1/2	1/2

a) Most glass crushers will accept 1 gal glass jars.

is used to puncture plastic containers in order to increase bale density with resultant shipping economies. The perforations also eliminate the need to remove bottle caps and improve baler efficiency since bales are easier to form. Ample storage must be provided for the perforated containers so that the baler may be efficiently utilized.

Baler

A baler is one of the most common pieces of processing equipment employed in a MRF. A diagram of a baler is presented in Figure 2-23. Balers are used for forming bales of newsprint, corrugated, high-grade paper, mixed paper, plastics, aluminum cans, and tin cans. These units are available with a wide range of levels of sophistication. Some balers are equipped for fully automatic operation while others demand a considerable amount of operator attention. If the design calls for the use of the same baler to bale more than one material, it is extremely important that the baler selected be specifically designed for that purpose. The market specifications which must be met should be determined before a baler is selected. Not all automatic tie devices are alike. The number and size of baling wires as well as the available wire tension must be adequate for the particular materials to be baled.

Figures 2-24 and 2-25 depict the typical range of dimensions for single ram and two ram balers respectively as commonly used in MRFs. Clearance should be provided for maintenance and accumulation of finished bales. Table 2-14 lists typical production rates for OCC versus

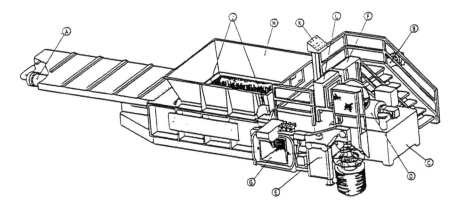

Figure 2-23. Baler.

Table 2-14. Typical OCC Production Rates* and Horsepower for Horizontal Balers.

TPH	HP
5-6	50
4-13	75
7-19	100
8-23	150
10-28	200
22-31	300

a) Production rates will vary due to bale dimensions, bale density, baler configuration, and other factors.

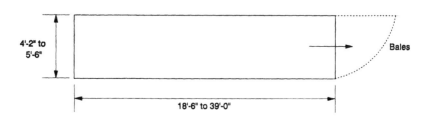

Figure 2-24. Typical range of dimensions for single ram balers.

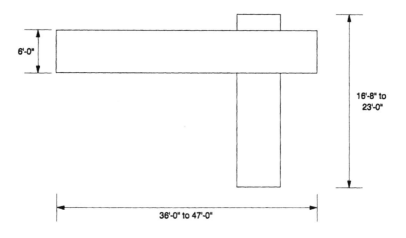

Figure 2-25. Typical range of dimension for two ram balers.

horsepower for horizontal balers commonly employed in MRFs. Typical dimensions, densities, and weights of bales for a variety of materials are given in Table 2-15. A considerable amount of research as well as test and evaluation of size reduction equipment has been carried out during the past 20 years. Some sources of information include References 6, 7, and 23 to 33.

2.3.10.2.4 Equipment for environmental control. In order to protect the health and safety of the work force as well as to gain the goodwill and to meet environmental requirements of the community in which the MRF is located, it often is necessary to provide environmental equipment above and beyond that which normally is supplied with the material handling, separation, and/or size reduction equipment. Title 29 of the Code of Federal Regulations, Part 1910, presents the Occupational Safety and Health Administration (OSHA) Standards which must be met to provide for the safety and health of the workers. Local and/or regional codes or legislation often address the environmental relationship of a facility within the community. In the planning and design phase of the facility it is wise to review those operations likely to cause distress to either the worker or to the community (or both) and seek ways in how to best ameliorate or eliminate the problems.

Dust Collection System

Shredding, granulating, crushing, baling, and screening generally are dust producing operations. Depending upon the severity (which often is a function of the volume of material handled) of the problem, the solution can vary anywhere from a simple dust mask for the worker, to individual dust collection at each of the dust producers to one or more centralized dust collection systems to serve the total needs of the facility. Dust collection systems include fans, ducting, cyclones, and baghouses.

Noise Suppression Devices

The majority of the equipment used in MRFs generate noise and/or dust. As is the case with dust problems, the solution to noise problems can be simple (e.g., hearing protection worn by the worker) or may require sound muffling equipment and/or sound proofing at specific work

Table 2-15. Typical Densities and Weights (40"x30"x62" Bales)* for Baled Materials.

Component	Baled Density (lb/cu ft)	Baled Weight (lb)
Corrugated	25 - 33	1200 - 1600
News	30 - 40	1450 - 1940
PET	24 - 32	1160 - 1540
Aluminum Cans	15 - 46	730 - 2230
Steel Cans	30 - 60	1450 - 2900
Solid Waste	38 - 54	1840 - 2610

* Bale sizes, volumes and weights may vary by baler manufacturer, model, mode of operation, moisture content, and other factors.

locations or throughout the building, or isolation of specific pieces of equipment.

Odor Control Equipment

Odor control is not generally a problem at a MRF unless the MRF is processing mixed MSW. Odors can often be reduced or eliminated by minimizing storage time of raw materials or product followed by frequent floor washdown. Other odor control technologies include:

- Improved dispersion
- Odor masking
- Wet scrubbing
- Carbon adsorption
- Catalytic incineration
- Thermal incineration

In severe odor conditions, multiple technologies may be required. Each technology may be accompanied by problems (in addition to capital and operating costs) of its own and indeed, the technology may not be acceptable to the control agency or to the complainants.

Heating, Ventilating, and Air Conditioning (HVAC)

The geographical location (and the associated climatic condition) of the facility has a major influence on the HVAC system required as does the very building design itself. Some MRFs incorporate enclosed sorting station rooms in which HVAC systems can be more effective than for open stations. Area heaters and ceiling and wall insulation may also be employed. Adequate ventilation must be provided to control fumes which may be generated by material handling vehicles, incidental hazardous incoming materials, and incidental welding operations.

2.3.10.3 Fixed Equipment Capacity

The process of MRF design should include that the manufacturer's rated capacity and maximum capacity, generally expressed in tons per hour (TPH) for conveying, separating, and processing equipment, be established and guaranteed. For equipment in a system in which there is no redundancy, it is wise to incorporate extra capacity in order to compensate for the inevitable downtimes. Alternatively, the equipment may be called upon to operate on an overtime basis.

Example:

A paper baler has the following characteristics:

    Rated Capacity: 25 TPH
    Maximum Capacity: 27.5 TPH

The baler will have the following schedule for normal operation:

    Number of hours per day: 8
    Number of days per week: 5

Assuming that the baler is out of service for repair for 8 hours during a one-week period, it is necessary to calculate the options for making up the loss in production. The expected production can be obtained by multiplying the rated capacity by the number of hours of normal operation.

    Expected production = 25 TPH x 40 hrs = 1,000 tons

The "actual" production, however, is calculated based on only 32 hours of operation. Thus:

Actual production = 25 TPH x 32 hrs = 800 tons

Consequently, there is a deficiency of production of 200 tons (1,000 tons - 800 tons).

The following options can be followed in order to makeup the deficiency:

Option 1

The baler could be operated for 7.3 hours of overtime at maximum capacity (27.5 TPH).

200 tons = 7.3 hrs x 27.5 TPH

Option 2

The baler could be operated for 8 hours at rated capacity (25 TPH).

200 tons = 8 hrs x 25 TPH

Option 3

The baler could be operated for 80 regular hours at maximum capacity.

200 tons = (80 hrs x 27.5 TPH) - (80 hrs x 25 TPH)
200 tons = 2,200 tons - 2,000 tons

2.3.10.4 Fixed Equipment Material Recovery Efficiencies

As discussed in subsection 2.3.5. and listed in Table 2-6, there are various factors which affect the recovery rate of materials. As shown in Table 2-16, the interaction of these factors result in a fairly broad range of material recovery efficiencies.

In each case, the low end of the efficiency range indicated in Table 2-16 may be reached when the feed rate is heavy and the time of exposure of the material to the separation device is minimal. Conversely, the higher recovery efficiencies may be realized at light feed rates (e.g., where a can or bottle is not buried in the waste stream) and the time of exposure of the material to the separation device is maximized.

2.3.10.5 Availability of Fixed Equipment

Availability is defined as the estimated portion of time that a particular piece of equipment is available to perform the work for which it is intended. This is a concept often overlooked in the equipment selection process. The concept of availability takes on special significance when the equipment in question is one of a series of machines as is generally the case in a processing system. For example, assume that a single processing line consists of 5 pieces of equipment served by 6 conveyors. Also assume that, for the purpose of illustrating the concept, the availability of each of the machines is 0.95 (i.e., each machine is expected to be down for repair, maintenance, pluggage clearance, power outage, etc., 5% of the time that it might otherwise be running). Lastly, assume that the availability of each of the conveyors is 0.99. Then, the availability of the total system (i.e., the process line) on a worst case basis (i.e., any given machine or conveyor is unavailable when all others are available), is:

Conveyors (99%) 6 x Machines (95%) 5 = 72.8%

In other words, a system using these machines and conveyors all in line in this manner for a 40 hour period would, on a worst-case basis, operate only 0.728 x 40 hours = 29.12 hours. The example is provided to underscore the importance of the concept and is not meant to suggest actual availabilities of specific equipment.

Equipment and system availability can be improved in various ways. Some of these are:

- The selection of proven equipment with a documented and validated history
- The selection of heavy duty equipment
- Proper system design which anticipates jam and pluggages, particularly at entrance, transfer, and discharge points and provides for their relief or elimination

- Trained operating personnel who understand the limitations of the equipment
- Trained maintenance personnel who can readily address downtime problems
- Preventative maintenance program
- Supply of spare parts with particular attention to long lead items
- Awareness, in the design phase, of the interrelationship of equipment so that the discharge from one machine is compatible with the operations of downstream equipment.

Table 2-16. Material Recovery Efficiencies for Separating Equipment.

Machine	Typical Range of Material Recovery Efficiencies (%)
Magnetic Separator (ferrous)	60 - 90
Eddy Current (aluminum)	60 - 90
Disc Screen	50 - 90
Trommel Screen	80 - 90
Vibrating Screen	60 - 90
Traveling Chain Curtain	60 - 90
Air Classifier	60 - 90

Source: CalRecovery, Inc.

2.3.10.6 Fixed Equipment Redundancy

Problems related to capacity and availability can be substantially reduced by providing multiple machines and/or processing lines. This concept is known as redundancy. Judicious use of redundancy in a design implies that if a machine or processing line is out of service for any reason, another machine or line can be brought into operation. Provision for redundancy, however, is often accompanied by a requirement for increased capital expenditure, not only for the duplicate equipment but also for the additional building space necessary to house that equipment.

A form of redundancy can be achieved by other less expensive means, including:

- Use of common parts. For example, standardizing belt widths, motor sizes, and other mechanical and electrical components will reduce the spare parts inventory yet allow ready repair of equipment.
- Multiple-use equipment. A paper baler, for example, may be equipped to handle plastics, tin, and/or aluminum.
- Use of diverters. For example, in anticipation of downtime of a glass crusher for flint glass, it may be feasible to temporarily divert that material to another glass crusher for processing.
- Markets may be available which suggest that redundancy in some equipment should be of minor importance. For example, there may be a market for PET in ground or baled form. However, even at some price reduction for the final product, it may be wise to plan on selling the baled product at the lower price rather than incur additional capital and operating costs which may be associated with the granulating process.

Redundancy is a very important concept in the design of MRFs. Redundancy is particularly important at points or sections of a system that are critical to the continuous operation of the plant. The implementation of redundancy must be carefully balanced with practicality and financial viability.

2.3.10.7 Sizing of Fixed Equipment

The considerations of recovery efficiency, capacity, availability, and redundancy discussed in the preceding section in addition to anticipated fluctuations in the daily quantities of materials received, the size of the tipping floor, the number of shifts planned for operating, budgetary constraints, and the degree of risk one is willing to accept, all influence the design and selection of individual pieces of fixed equipment.

It must be emphasized that average daily tonnages calculated by simply dividing the annual tonnage by the number of operating days (see subsection 2.3.2., Mass Balance) can be quite misleading when designing and selecting equipment.

If one ignores budgetary constraints, a capacity safety factor or multiplier, ranging from 1 to 2 on the maximum daily tonnages of materials anticipated, should be considered. For example, a multiplier of unity would be reasonable for equipment sizing if the facility was designed with total redundancy (with each piece of equipment capable of handling the full load), high equipment availability (proven equipment and systems), single shift operation (with the option of operating a second shift), and a relatively consistent flow of materials. A multiplier of two would be reasonable for equipment sizing if the facility was designed with little or no redundancy, low availability due to the positioning of many pieces of equipment in series, two shift scheduled operation, and large fluctuations in inflow of materials. For equipment employed in the average MRF, a multiplier of 1.25 to 1.5 generally is used. The unique concerns relating to the sizing of sorting conveyors will be discussed in subsection 2.3.12.2 entitled Manual Sorting Rates and Efficiencies.

2.3.10.8 Maintenance of Fixed Equipment

Early in the design phase of a MRF, consideration should be given to providing sufficient access to the fixed equipment for the maintenance and repair work required to keep the Facility operational. Preparation of preliminary maintenance procedures (preferably with the assistance of the equipment supplier) similar to those examples illustrated in Appendix 3 for belt conveyors, magnetic separator, trommel screen, can flattener, and baler serve not only to identify and evaluate the amount and quality of maintenance required but also to alert the designer to those equipment components to which access must be provided.

2.3.11 Rolling Equipment

The review and selection process of rolling equipment for use in a MRF employs much of the same rationale as that outlined for the review and selection process for fixed equipment. The following observations concern some special considerations associated with rolling equipment.

2.3.11.1 Rolling Equipment Commonly Found in a MRF

- Bins
- Containers
- Floor scrubber
- Forklift
- Front-end loader
- Manulift
- Skid steer loader

- Steam cleaner
- Vacuum/Sweeper/Magnetic pick-up
- Yard tractor

2.3.11.2 Rolling Equipment Capacity

Rolling equipment (most of which is material handling equipment) must, of course, be adequate to perform the tasks required to feed the plant, perform intermediate material transfers, and to load out the products. Equipment must be selected of adequate power, speed, and size to handle the tonnages anticipated. If the equipment is too small, the productive capacity of the entire plant can be adversely impacted. It is also possible for the equipment to be too large for the plant in that there may not be enough room to maneuver.

The information presented in Table 2-17 is provided as a guide in the selection of an appropriate bucket size for a front-end loader handling the materials generally processed in a MRF.

2.3.11.3 Availability of Rolling Equipment

Rolling equipment should be considered as an integral part of the process line of a MRF. Downtime associated with rolling equipment which delivers material to an infeed conveyor, transfers material to or from various processes, or loads the product into or onto outgoing trucks, trailers, etc., affects the overall plant availability just as does fixed equipment downtime. The same list of considerations provided under that for fixed equipment apply to rolling equipment.

2.3.11.4 Rolling Equipment Redundancy

The requirement for redundancy in rolling equipment is not as severe as that for fixed equipment. Standard models of various pieces of rolling equipment are often readily available for temporary or emergency use from a local dealership. Often, as part of maintenance/service contract, a rolling equipment dealer will make available a replacement unit in the event that a particular machine must undergo extensive repair. Additionally, various attachments to basic machines may provide a degree of redundancy through multi-purpose use.

Table 2-17. Examples of Front-end Loader Capacities.

Component	Average Loose Bulk Density[a] (lb/cu yd)	Approximate TPH per cu yd Bucket Capacity of Front-End Loader[b]
Whole Containers		
Glass	500	7.50
Plastics	65	1.00
Aluminum	65	1.00
News	170	2.60
OCC	50	0.75
Loose Refuse	150	2.30
Refuse after dumping from compactor truck	375	5.60

a) The values used are averages of a range of available data for each component.
b) For other front-end loader capacities, multiply the relative bucket size.
   Bucket size = 2-1/4 cu yd. Glass = 7.5 TPH/cu yd x 2.25 = 16.9 TPH

2.3.11.5 Rolling Equipment Selection

In the review and selection process of individual items of rolling equipment, just as for fixed equipment, it should be recognized that these items must not only compatibly interrelate with one another, but also with the manner in which the raw material is to be received, the in-process material transferred, and the product loaded for shipment. Special care should be given as to whether or not the vehicle is to be used exclusively indoors or outdoors, or both, particularly in regard to exhaust fume generation.

2.3.12 Human Factors

The purpose of this section is to explore a few of the psychological and physical relationships that arise as workers interact with machinery in a MRF environment.

2.3.12.1 Staffing Requirements

Whether a MRF utilizes a low or high technology system configuration or some intermediate system, there is a need for the employment of manual laborers. In another section of this Manual, job descriptions, employee relations, health and safety, and other topics will be discussed. The information presented in Table 2-18 is provided as a guide to the size and make-up of the work force in MRFs of various throughputs.

2.3.12.2 Manual Sorting Rates and Efficiencies

The ranges of manual sorting rates and efficiencies for various materials are presented in Table 2-19. In a mix of materials, such as OCC and newspaper, higher sorting efficiencies will generally be achieved by manually removing ("positively sorting") the lesser quantities of material from the greater. As shown in Figure 2-5, in Sorting Station #1 residue at a rate of 4.4 TPD and OCC at a rate of about 9.8 TPD are positively sorted from the incoming mix of 44 TPD. On the other hand, about 29.77 TPD of newspaper are permitted to pass through the sorting station untouched (i.e., "negatively sorted"). With reference to Sorting Station #2 in Figure 2-6, mixed broken glass would be negatively sorted even though it represents a lower throughput than either green or clear glass since the broken glass would be more difficult to manually extract from the glass stream than whole glass containers.

Sorting stations should be arranged so that the sorters are not competing with one another for the same item. Some designers accomplish this by positioning the sorters on only one side of the sorting belt and by assigning specific materials to be handled by each sorter. Other designers locate the sorters on both sides of the sorting belt. In this particular situation, the sorting positions are staggered along the belt length in order to avoid competition by more than one sorter over the same item. Recommended maximum widths for sorting belt selected to minimize personnel fatigue and consequent loss of efficiency are given in Table 2-20. The working height of the sorting belt should be between 36 in. and 42 in. from the platform level. A working height of 42 in. allows for the installation of temporary risers for shorter workers.

Sorting belts should be outfitted with variable speed devices capable of controlling the belt speed between 0 and 100 FPM. The higher belt speeds would be utilized under conditions where most or all of the material is anticipated to be negatively sorted. For average sorting conditions for both paper sorting and commingled container sorting, a maximum belt speed of 30 FPM is considered appropriate. Sorting rates and manual material recovery efficiencies may be further enhanced by

Table 2-18. Approximate Range of Staffing Requirements for Material Recovery Facilities.

Personnel	Tons per Week			
	500	1,000	1,500	2,000
Office				
Plant Manager	1	1	1	1
Scalemaster/Bookkeeper	1	1	1	1
Clerk	0 - 1	0 - 2	2 - 3	2 - 3
Janitor	0	0	0	0
Plant				
Foreman/Machine Operator	1 - 2	2 - 3	3 - 4	3 - 4
Sorters	13 - 25	16 - 27	19 - 32	25 - 38
Forklift/FEL Operators	2 - 3	3 - 4	4 - 5	5 - 6
Maintenance	1	2	3	4
TOTAL	**19 - 34**	**26 - 40**	**33 - 49**	**42 - 58**

Source: CalRecovery, Inc.

Table 2-19. Manual Sorting Rates and Efficiencies

Material	Approximate Ranges			
	Containers/lb	Containers/Minute/Sorter	Lb/Hr/Sorter[a]	Recovery Efficiency (%)
Newspaper	----------	------------	1,500 - 10,000	60 - 95
Corrugated	----------	------------	1,500 - 10,000	60 - 95
Glass (mixed/whole)	1.5 - 3.0	30 - 60	900 - 1,800	70 - 95
Glass (by color)	1.5 - 3.0	15 - 30	450 - 900	80 - 95
Plastic (PET, HDPE)	4.5 - 9.0	30 - 60	300 - 600	80 - 95
Aluminum (from plastic)	22.5 - 27	30 - 60	100 - 120	80 - 95

a) Based on average sorting rates (containers/minute/sorter).

Table 2-20. Recommended Maximum Sorting Belt Widths

	Recommended Maximum Belt Width (in).	
Sorting Process	Sorting Stations on One Side	Sorting Stations on Both Sides
Paper (OCC or ONP)	42	72
Commingled Containers	30	48

providing the sorting area with complete environmental control (i.e., heating, ventilation, and air conditioning). This approach will also reduce personnel exposure to process noise and dust.

Sample Calculations (Paper)

Refer to Figure 2-5, Paper Line - Sorting Station #1
Incoming paper = 44 TPD (5.5 TPH)
Design capacity = 1.5 x 5.5 TPH = 8.25 TPH

To find combined density:

Newspaper (6.19 lb/cu ft) x 29.77 TPD = 184.3
OCC (1.87 lb/cu ft) x 9.83 TPD = 18.4
Residue (150 lb per cu yd/27) x 4.5 TPD/44 TPD = 24.5/227.2
Average density = 227.2/44 = 5.16 lb/cu ft

Capacities for flat belts (in cu ft/hr) at a speed of 100 FPM are presented in Table 2-21.

Conclusion

In the example chosen, a width of 72 in. for the sorting belt at a speed of 30 FPM is an option which would provide sufficient capacity to accommodate material surges of 50% of the nominal feed rate. Consequently, lacking budgetary constraints, this belt is the recommended choice for this particular application.

Table 2-21. Flat Belt Capacity[a]

Belt Width (in.)	Capacity at 100 FPM (cu ft/hr) Surcharge Angle					
	5°	10°	15°	20°	25°	30°
18	120	246	372	498	630	762
24	234	465	702	942	1188	1446
30	378	756	1137	1527	1926	2340
36	552	1112	1677	2253	2844	3450
42	768	1542	2322	3120	3936	4776
48	1014	2037	3072	4128	5208	6318
54	1296	2604	3924	5274	6654	8076
60	1614	3240	4884	6560	8280	10050
72	2352	4722	7116	9558	12060	14640
84	3228	6480	9768	13116	16548	20091
96	4243	8514	12834	17238	21750	26406

a) Standard Edge Distance = 0.55b + 0.9 in. Adapted from CEMA "Belt Conveyors for Bulk Materials."

Table 2-21. From Table 2-21 and Surcharge Angle = 30°

	Belt Width (in.)				
	42	48	54	60	72
Cu ft/hr at 100 FPM	4776	6318	8076	10050	14640
Cu ft/hr at 30 FPM	1433	1895	2423	3015	4392
Lb/hr at 30 FPM and 5.16 lb/cu ft.	7394	9778	12503	15557	22663
TPH at 30 FPM and 5.16 lb/cu ft.	3.7	4.9	6.3	7.8	11.3

Safety factors based on 5.5 TPH nominal feed rate:

   54 in. belt with 6.3 TPH/5.5 TPH = 1.1
   60 in. belt width 7.8 TPH/5.5 TPH = 1.4
   72 in. belt width 11.3 TPH/5.5 TPH = 2.1

Table 2-21. From Table 2-21 and Surcharge Angle = 5°

	Belt Width (in.)				
	24	30	36	42	48
Cu ft/hr at 100 FPM	234	378	552	768	1014
Cu ft/hr at 30 FPM	70	113	166	230	304
Lb/hr at 30 FPM and 15.76 lb/cu ft	1103	1781	2616	3625	4791
TPH at 30 FPM and 15.76 lb/cu ft	0.6	0.9	1.3	1.8	2.4

Safety factor based on 1.77 TPH feedrate:

    36 in. belt width 1.3 TPH/1.77 TPH = 0.7
    42 in. belt width 1.8 TPH/1.77 TPH = 1.0
    48 in. belt width 2.4 TPH/1.77 TPH = 1.36

Sample Calculations (Commingled Containers)

Refer to Figure 2-6, Commingled Container Line -- Sorting Station #2
Incoming containers = 14.18 TPD (1.77 TPH)
Design capacity = 1.5 x 1.77 = 2.66 TPH

To find combined density:

    Glass (18.45 lb/cu ft) x 11.35 TPD = 209.4
    Ferrous (4.9 lb/cu ft) x 0.62 TPD = 3.0
    Aluminum (2.36 lb/cu ft) x 0.10 TPD = 0.2
    Plastic (2.37 lb/cu ft) x 0.23 TPD = 0.5
    Residue (150 lb per cu yd/27) x 1.88 TPD/14.18 TPD =
       10.4/223.5
    Average density = 223.5/14.18 = 15.76 lb/cu ft

Conclusion

In the example chosen, a width of 48 in. for the sorting belt operating at 30 FPM is the only possibility which would provide sufficient capacity to accommodate any material surges (and that would be approximately 36% over the nominal feed rate). It is not suggested that a wider belt be used since that would reduce worker efficiency. If necessary, for short periods of time, the belt could be operated at a higher speed (2.66 TPH/2.4 TPH x 30 FPM = 33 FPM) in order to reach a 50% surge capacity.

2.3.12.3 Psychological Factors

In the long list of services and processes provided by individuals and organizations in the communities which make up our society, few may be regarded as more beneficial to and necessary for our society than those associated with a MRF.

Much can be done to enhance the status of the manual laborer both in the eyes of the public as well as in his or her own eyes. They include, among others:

- Conducting an active and continuing public relations campaign citing the important contribution a MRF makes in improving the quality of life
- Designing and building a MRF which is aesthetically pleasing both to the visitor and to the worker
- Developing a sense of pride and accomplishment in the mind of each worker for a difficult task well done
- Maintaining the MRF in such a manner as to make it as pleasant a place as possible in which to work

2.3.12.4 Physical Factors

As is true with many processing and/or manufacturing plants, workers in a MRF must interact with both fixed and rolling equipment on a continual basis. In a MRF, the incoming materials, particularly bottles (broken glass) and cans (sharp, ragged edges), present physical dangers to the worker as does the equipment employed in handling and processing that material.

In addition to the attention which must be paid to providing each worker with safety clothing and equipment and otherwise adhering to the general industrial safety practices (OSHA), there are a few special precautions to observe in the design and operation of a MRF. They include, among others:

- Incorporating a system in which the worker monitors the machine. This is to ensure that the machine operates as intended and is not overloaded;
- Incorporating a system in which the machine monitors the worker. This is to ensure that should the worker, for whatever reason, not perform the task as intended, the machine will issue a warning or shut down;
- Adopting an operating philosophy that the worker is not in competition with the machine, but rather that the worker and machine complement one another in order to best perform the task;
- Designing the work stations in such a manner as to limit the physical exertion and awkward bending, stretching, lifting, and moving required to perform the task;
- Arranging equipment controls in a simple and consistent manner from machine to machine to reduce the chance of operator error; and
- Recognizing the probability of fatigue or boredom because of the routine nature of the tasks and adjusting working schedules and/or task assignments accordingly.

2.3.12.5 Employment Opportunities

In communities of high or chronic unemployment (particularly of unskilled laborers), MRFs present an opportunity to alleviate that condition. The MRFs also provide an opportunity for the employment of part-time seasonal workers typically required in resort areas.

2.3.13 Acceptable Waste

Acceptable waste may be defined as that material which the MRF is designed to receive and process for the markets identified. For the Basic MRF as described in subsection 2.2.1, acceptable waste is identified as source separated materials arriving at the facility in two waste streams, i.e., paper and commingled containers. Variations from the Basic MRF which broaden the list of acceptable waste are discussed in subsection 2.2.2.

The quality of the incoming waste materials is highly dependent upon the understanding, co-operation, and participation of the public. In order to increase the probability of the facility receiving acceptable waste, it is wise to publish a list, not only of acceptable waste and how to prepare it, but also of waste that is unacceptable. One such list is shown on Table 2-22.

2.3.14 Raw Material Storage

Ideally incoming waste will be tipped directly onto an infeed conveyor and processed as received obviating the need for raw material storage. The advantages in so doing include a lesser requirement for floor space, a lesser requirement for material handling rolling equipment on the receiving floor, and a shorter total residence time for the material from receipt to shipment.

Unfortunately, things do not always run that smoothly. In order for the incoming waste to be tipped directly onto an infeed conveyor some substantial co-ordination and control is necessary between the MRF and the collection operation so that collection vehicles are not lined up interminably waiting to deliver their loads. Also, unless there is some redundancy in the provisions for receiving and processing, system outage can also result in lengthy queues of vehicles waiting to unload or to be diverted to the landfill.

The design of most MRFs incorporate sufficient storage area to accommodate the equivalent of at least one day's supply of raw material. Several factors influence the decision regarding the amount of floor space to allocate to raw material. They include:

- Redundancy. A facility with redundant processing systems has less need for raw material storage space
- Processing vs receiving hours. A facility open to receipt of raw material outside of scheduled processing times must provide sufficient storage capacity for the raw material. In the case where scheduled processing takes place (e.g., a second shift) beyond the MRF receiving hours, raw material storage is also necessary in order to provide the material to process
- Local regulations: In many localities restrictions are placed upon the number of vehicles which may queue up to unload. Adequate raw material storage space must be provided to prevent this condition from occurring
- Vehicles vs tipping floor configuration. The mere provision of floor space for the storage of raw materials may not totally address the problems discussed above. Care must be taken that the collection vehicles can gain ready access to the tipping floor,

quickly unload, and depart with a minimum of interference with other vehicles and/or the front-end loader(s) on the tipping floor.

Table 2-22. Materials to be Collected for Recycling Newspaper, White Office Paper, Corrugated Cardboard, Glass, Tin Cans, Aluminum, and Plastic[a]

Acceptable (Common Names)	Non-Acceptable	Preparation
Paper Newsprint White office paper	Glossy paper Magazines Phone books Colored office paper	Remove any tape, rubber bands, or staples
Cardboard Corrugated	Any waxed cardboard (i.e., milk cartons) Any corrugated contaminated with food or other waste	Flatten corrugated
Glass Bottles (any color) Jars (any color)	Plate glass (window) Light bulbs Drinking glasses Ceramics of any kind	Do not break Rinse Remove tops, rings and caps May leave paper labels
Aluminum Aluminum beverage cans, foil, aluminum pie plates	Construction aluminum	Rinse and clean
Tin Cans All tin cans	Unwashed cans	Rinse can Remove label Do not need to remove both ends or flatten
Plastic Only consumer (i.e., high density polyethylene (HDPE), shampoo bottles, detergent bottles, milk and water bottles, oil, anti-freeze containers) PET = beverage containers	Any brittle plastics (i.e., cottage cheese containers) Film (i.e., plastic bags) Ketchup bottles Industrial plastic	Remove caps and rings Rinse container Flatten if possible

a) For example purposes only

2.3.15 Product Storage

The amount and location (i.e., indoors or outdoors) of space allocated to product storage is influenced, in great degree, by the markets.
Paper products may be stored indoors in bales, loaded loose into compactor type transport vehicles, or baled and loaded into trailers or rail cars. Often the market will dictate on how the product is to be shipped. Aluminum cans, tin cans and bimetal cans may be shipped loose, flattened or otherwise size reduced and shipped.   The market specifications will also influence whether or not the products may be stored outdoors pending shipment. The finished forms of other products as well have been discussed in subsection 2.2.1. and included in Tables 2-2 and 2-3. In any case, sufficient space must be allocated indoors or outdoors in order to make an economic shipment of the product to the market.

2.3.16 Building

The MRF building design should be a joint effort on the parts of the process engineer, structural engineer, and the architect. The design will necessarily be influenced by the site conditions and anticipated traffic patterns.   Clear, wide bays utilizing a minimum number of interior columns are preferred in order to present the least possible interference with trucks and other rolling equipment. A high bay tipping floor is often a requirement for the accommodation and dumping of raw material. Similarly, wide high bay doors are desired to minimize the possibility of interference with tipping vehicles.

Consideration should be made in the design of the building for the possible future expansion of the facility to handle greater quantities and/or an increased variety of raw materials. The building should also be viewed as a tool for the mitigation of any noise, dust, litter, and odor that might otherwise adversely impact upon the surrounding neighborhood. Enclosed, well illuminated sorting rooms with properly designed HVAC systems will assist in maintaining a high level of productivity and worker morale. Obviously, all Building, Fire, and Safety Codes must be adhered to.

2.4 MRF MANAGEMENT

2.4.1. Organization

A nationwide survey of MRFs (Table 2-23) gives a numerical breakdown of employees by management and nonmanagement categories, and by size of facilities. The total number of employees per existing facility averages about 19. Planned installations at the time of the survey, showed a higher workforce, approximately 26; however, these planned facilities are larger in design capacity than current plants.

Data relating the ratio of nonmanagment employees to design capacity are given in Table 2-24. This ratio can provide an indication of operational effectiveness, as labor costs are a significant part of total operation and maintenance costs. However, lower labor requirements normally result from the use of fairly sophisticated equipment. The increased capital costs must then be equated with lower costs before a judgement on operational efficiency can be made. In Table 2-23, the ratios of employees to different categories of MRFs is also given. These types include: plants processing 1 to 99 TPD, and greater than 100 TPD; planned and existing facilities; and "low" and "high-tech" plants.

2.4.2 Operating Schedules

The majority of MRFs surveyed (GAA, 1990) processed materials on a 5-day per week basis. The mean figure, as shown in Table 2-24, actually is 5.23 days per week with a range of 4.0 to 6.5 days per week.

Most of the MRFs surveyed operated one shift per day; some scheduled two or three shifts. The mean value for all facilities was 1.16 shifts per day. The length of a shift was 8 hours at nearly all planned and existing facilities; however, a small number of existing plants had shifts ranging from 4 to 10 hours. The average number of days that the MRFs were in operation varied between 208 and 338; the average number of days was approximately 266.

It is important to note that the schedule of operations for any facility will depend on locally defined conditions. These conditions would include collection schedules, throughput of the facility, capacity of the facility, etc.

Table 2-23. Number of Full-time Equivalent (FTE) Employees

Sample	Mean	Sum	Standard Deviation	Minimum	Maximum	N
Total Employees						
All Facilities	21.67	1,538	23.37	1.50	165.00	71*
Planned Facilities	25.69	796	28.56	8.00	165.00	31
Existing Facilities	18.55	742	18.17	1.50	92.00	40
Management						
All Facilities	2.87	204	2.96	0.10	16.00	71
Planned Facilities	3.56	110	3.57	1.00	16.00	31
Existing Facilities	2.34	94	2.28	0.10	12.00	40
Non-Management						
All Facilities	18.79	1,334	20.94	1.00	150.00	71
Planned Facilities	22.13	686	25.69	7.00	150.00	31
Existing Facilities	16.21	648	16.22	1.00	80.00	40
Ratio of Non-Management Employees: Design Capacity (Tons Per Day)						
All Facilities	0.272	-	0.442	0.035	3.500	71
1 to 99 TPD	0.385	-	0.578	0.044	3.500	38
100 + TPD	0.142	-	0.142	0.035	0.468	33
Planned	0.159	-	0.155	0.035	0.900	31
Existing	0.360	-	0.561	0.044	3.500	40
Low-Tech	0.313	-	0.582	0.058	3.500	38*
High-Tech	0.210	-	0.130	0.035	0.535	31

* No information was available from 33 planned MRFs with regard to number of employees (management or non-management); an additional two projects did not furnish data with respect to degree of mechanization.

Table 2-24. Operating Schedules of MRFs

Sample	Mean	Standard Deviation	Minimum	Maximum	N
Days of Operation Per Week					
All Facilities	5.23	0.42	4.00	6.50	99*
Planned Facilities	5.24	0.40	5.00	6.50	59
Existing Facilities	5.21	0.45	4.00	6.00	40

* No information was available from five planned MRFs with regard to days of plant operation per week.

Sample	Mean	Standard Deviation	Minimum	Maximum	N
Shifts Per Day					
All Facilities	1.16	0.41	1.00	3.00	98*
Planned Facilities	1.13	0.42	1.00	3.00	58
Existing Facilities	1.19	0.39	1.00	2.00	40

* No information was available from six planned MRFs with regard to the number of shifts per day.

Sample	Mean	Standard Deviation	Minimum	Maximum	N
Hours Per Shift					
All Facilities	8.00	0.50	4.00	10.00	98*
Planned Facilities	8.00	0.00	8.00	8.00	58
Existing Facilities	8.00	0.50	4.00	10.00	40

* No information was available from six planned MRFs with regard to hours per shift.

Sample	Mean	Standard Deviation	Minimum	Maximum	N
Days of Operation Per Year					
All Facilities	266.51	21.29	208.00	338.00	98*
Planned Facilities	265.88	21.69	250.00	338.00	58
Existing Facilities	267.43	20.94	208.00	312.00	40

* No information was available from six planned MRFs with regard to the number of days of plant operation per year.

2.4.3 Job Descriptions

A variety of skills are required for personnel operating a MRF. Descriptions of jobs to be carried out at a MRF are discussed in the following paragraphs.

2.4.3.1 Plant Manager

The plant manager works under the general supervision of an operations vice president. The plant manager directs and coordinates, through subordinate supervisory personnel, all activities concerned with production of end products from the recyclables. The manager will confer with management staff at the corporate level to ensure achievement of established production and $/annual capacityquality control standards, development of and compliance with cost controls, development of operational budget, and maintenance of the safety plan.

2.4.3.2 Foreman

The foreman works under the direct supervision of the plant manager. The foreman is responsible for the daily production of end products in specified quantity and quality on both the mixed recyclables and paper processing lines. The foreman will conduct start-up and close-down procedures before and after his shift and ensure that proper maintenance procedures are followed by the employees under his supervision. Other responsibilities include: inspecting load-out of materials; ensuring that all work stations are maintained in a clean and orderly manner; and verifying that all employees are furnished with appropriate safety apparel and equipment.

2.4.3.3 Maintenance Mechanic

The maintenance mechanic repairs and maintains, in accordance with diagrams, sketches, operations manuals, training programs, and manufacturer's specifications, all machinery and electrical equipment relating to processing. The maintenance mechanic is also responsible for performing maintenance checks before and after operations, as well as initiating purchase orders for necessary parts and supervising general factory workers in cleaning and preventive maintenance tasks on individually assigned equipment. The senior maintenance mechanic reports directly to the foreman.

2.4.3.4 Equipment Operators

The equipment operator is responsible for movement and transfer of recyclables. Each operator has a complete material inspection and quality control. The operators on both lines are responsible for properly loading material to assure a fully charged receiving pit and a well-mixed load, and to densely and evenly load bales of processed material onto transfer trailers. One equipment operator is responsible for facilitating bailer-to-trailer loadout of processed steel, aluminum, and plastic. All rolling stock operators and plant personnel are cross-trained for versatility and plan efficiency.

2.4.3.5 General Factor Workers

General factory workers are responsible for color sorting glass and separating HDPE and PET plastics. All general factory workers are trained to ensure that contaminant-sensitive material (e.g., glass) is free of deleterious foreign matter such as ceramics, plate glass, and porcelain. General factory workers are trained to recognize nonrecyclable material and inform the foreman if any potentially damaging or hazardous items are found in the process flow. These are the only active sorters on the processing line.

2.4.3.6 Quality Assurance Inspectors

Quality assurance inspectors staff the mixed recyclables line at the inspection station. The inspector is responsible for examining the material infeed for contaminants and nonrecyclable materials. A dedicated inspector--not sorters preoccupied with maintaining end-product purity--is necessary to ensure the removal of these reject materials.

2.4.3.7 Administrative Assistant

The administrating assistant works under the direct supervision of the plant manager. Duties of the administrative manager include: "front office" tasks such as answering telephones and reception; preparing and submitting all required reports such as material shipments, and personnel record keeping.

2.4.4 Health and Safety Considerations

It behooves any employee of a MRF to be alert to potential health and safety problems associated with the workplace environment and the waste stream processed. There are physical danger inherent in the commingled recyclables or MSW, such as broken glass, sharp metals, etc. There are also potential environmental and medical dangers, particularly in raw MSW, blowing dust, etc.

Workplace dangers are also present at a MRF. Mobile equipment such as fork lifts, front-end loaders, and delivery trucks are heavily utilized; common sense safety procedures must be followed. Further, the nature of a MRF processing line requires that certain functions be carried out at elevated heights. With this in mind, there are steps to climb, sorting stations to tend, etc. Care must be exercised in getting to and from the work station, as well as while working. Safety helmets are a must, as a high probability exists that objects will fall from an elevated station from time to time.

Good safety practices are needed at any MRF. This necessitates a well-managed safety training program to inform the employee as to what constitutes "working safely;" this is a fundamental management responsibility.

2.5 MRF ECONOMIC ANALYSIS

2.5.1 Introduction

The purpose of this section is to present a range of capital and operating costs for material recovery facilities (MRFs). The costs for the facilities are presented in two forms: unit costs, such as dollars per ton per day ($/TPD), and in total cost for throughput capacities between 10 TPD and 500 TPD. A range of throughput capacities has been used to reflect any resultant economies of scale. A range of costs is presented in order to account for variations in both engineering design and in capital and operating costs, and to accommodate the wide variety of specific conditions that apply to MRF projects.

2.5.2 Composition of Recyclables

In order to perform the cost analysis for the facility, a composition of recyclable materials has been assumed. The assumed composition of recyclables expected to enter the facility is presented in Table 2-25. Furthermore, it has also been assumed that commingled paper will arrive into the facility separated from commingled containers (aluminum, steel, plastic and glass). This coincides with the material flow assumptions presented in Figures 2-5 and 2-6.

2.5.3 Capital Costs

2.5.3.1 Facility Construction Costs

Estimated capital costs have been developed for both facility construction and for equipment. Ranges for unit capital costs for five major construction categories are presented in Table 2-26. The difference between low and high cost ranges include project-specific conditions such as subsurface conditions, local topography, structural materials used for building construction (e.g., steel or concrete) and local building code requirements. Typical floor area requirements for the major sections of a MRF are presented in Table 2-27 as a function of throughput capacity. As indicated in the table, primary variables are the amount of tipping floor and storage capacity desired for processed recyclables. A general rule is to maintain sufficient tipping floor capacity to accommodate a reasonable "worst-case" unscheduled maintenance event and enough storage capacity for one to two unit truckloads (about 20 tons/truck) for each material processed.

The unit cost elements given in Tables 2-26 and 2-27 have been combined in Table 2-28 to present total and unit construction costs as a function of capacity. As shown in the table, in the case of facilities having a capacity in the range of 10 TPD to 500 TPD, the unit costs decrease as capacity increases.

2.5.3.2 Equipment Costs

Table 2-29 presents a range of typical unit equipment costs based upon the throughput capacity of the MRF. Similar to construction costs of the facility, the unit costs for the equipment decrease as capacity increases. Reasons for the decrease in unit costs include price reductions generally received from vendors for large purchases and economies of scale obtained when producing larger pieces of equipment, at least in the range of facility capacities considered herein. Table 2-30 presents total equipment costs by throughput capacity. The data in the table also show a summary of equipment unit costs.

Table 2-25. Assumed Recyclables Composition[a]

Material	Percent by Weight
Newspaper	33
Mixed Paper	41
TOTAL PAPER	74
Glass Bottles	19
Tin Cans	4
Aluminum Cans	1
PET & HDPE Containers	2
TOTAL COMMINGLED CONTAINERS	26
TOTAL	100

A) Recyclables are assumed to arrive at the MRF as commingled paper and commingled containers.

Table 2-26. Typical MRF Construction Costs[a] ($/sq ft Floor Area)

Item	Low	High	Average	Cost Segments
Site Work	$3.00	$10.00	$6.50	Excavation Grading Paving Landscaping Weight Scale
Utilities	$1.00	$2.00	$1.50	Electrical Water Sewage
Structures	$20.00	$40.00	$30.00	Concrete Structural Doors Indoor Utilities Fire Control Lighting
General Conditions b)	$1.00	$3.00	$2.00	Bonds Building Permit Mobilization
Contingency c)	$2.50	$5.50	$4.00	
TOTAL	$27.50	$60.50	$44.00	

a) Excludes engineering fee. See Table 2-29.
b) Equal to 5% of other construction costs.
c) Equal to 10% of other construction costs.

Table 2-27. Typical MRF Floor Area Requirements by Throughput Capacity (Sq. Ft.)[a]

Area Use	Capacity (TDP)		
	10	100	500
Tipping Floor b)			
2 Day Capacity	3,000	7,500	30,000
3 Day Capacity	3,000	11,250	45,000
Processing	6,000	20,000	50,000
Storage c)			
7 Day Capacity		8,750	35,000
14 Day Capacity	1,750	17,500	
28 Day Capacity	3,500		
TOTAL - Low	10,750	36,250	115,000
TOTAL - High	12,500	48,750	130,000
TOTAL - Average	11,625	42,500	122,500
FT2/TPD - Low	1,075	363	230
FT2/TPD - High	1,250	488	260
FT2/TPD - Average	1,163	426	245

a) Except as noted.
b) Assumes a density of 300 lb/cu yd, piled 12 feet high and a maneuvering factor of 2.5 for 10 to 100 TPD and 2 for 300 to 500 TPD
c) Assumed a processed material density of 800 lb/cu yd and maneuvering factors equal to those of the tipping fee.

Table 2-28. Estimated Construction Cost Range by Throughput Capacity[a]

Capacity (TPD)	Absolute Costs			Unit Costs		
	Low Cost (Total $)	High Cost (Total $)	Average Cost (Total $)	Low Cost ($/TPD)	High Cost ($/TPD)	Average Cost ($/TPD)
10	$295,625	$756,250	$511,500	$29,563	$75,625	$51,150
100	$996,875	$2,949,375	$1,870,000	$9,969	$29,494	$18,700
500	$3,162,500	$7,865,000	$5,390,000	$6,325	$15,730	$10,780

a) Costs based upon unit construction costs presented in Table 2-24 and floor area requirements shown in Table 2-25.

Table 2-29. Typical Unit Equipment Costs[a]

Equipment Item	10 to 100 TPD			500 TPD		
	Low	High	Average	Low	High	Average
SORTING SYSTEM						
Misc Conveyors	$1,500	$2,500	$2,000	$1,500	$2,000	$1,750
Sort Conveyors	$800	$1,200	$1,000	$800	$1,200	$1,000
Sort Platforms	$1,000	$2,000	$1,500	$500	$1,000	$750
Trommel Screens	$200	$500	$350	$100	$300	$200
Magnet Separators	$500	$1,000	$750	$300	$500	$400
PROCESSING SYSTEM						
Balers b)						
Paper	$2,500	$3,500	$3,000	$1,500	$2,000	$1,750
PET	(c)	(c)	(c)	$8,000	$12,000	$10,000
Metals	$40,000	$50,000	$45,000	$7,500	$10,000	$8,750
HDPE Granulators b)	$45,000	$70,000	$57,500	$30,000	$45,000	$37,500
Glass Crushers b)	$1,000	$2,500	$1,750	$1,000	$2,500	$1,750
ROLLING STOCK	$2,000	$2,500	$2,250	$700	$1,000	$850
INSTALLATION	10%	10%	10%	8%	8%	8%
CONTINGENCY	10%	10%	10%	10%	10%	10%

a) Unit costs for conveyors and platforms expressed as $/ft length. All others cost as $/TPD.
b) Unit costs are expressed at $/TPD of capacity for each listed material.
c) PET baled using the paper baler.

Table 2-30. Estimated Equipment Cost by Throughput Capacity

Equipment Item	10 TPD			100 TPD			500 TPD		
	Low	High	Average	Low	High	Average	Low	High	Average
SORTING SYSTEM									
Misc Conveyors	$75,000	$125,000	$100,000	$300,000	$500,000	$400,000	$750,000	$1,000,000	$875,000
Sort Conveyors	$32,000	$48,000	$40,000	$320,000	$480,000	$400,000	$800,000	$1,200,000	$1,000,000
Sort Platforms	$40,000	$80,000	$60,000	$400,000	$800,000	$600,000	$500,000	$1,000,000	$750,000
Trommel Screens	$2,000	$5,000	$3,500	$20,000	$50,000	$35,000	$50,000	$150,000	$100,000
Magnet/Eddy Seps	$6,000	$10,000	$7,500	$50,000	$100,000	$75,000	$150,000	$250,000	$200,000
SORTING TOTAL	$164,000	$268,000	$211,000	$1,090,000	$1,930,000	$1,510,000	$2,250,000	$3,600,000	$2,825,000
PROCESSING SYSTEM									
Balers									
Paper	$16,748	$23,447	$20,097	$167,475	$234,465	$200,970	$602,425	$669,426	$586,163
PET	$0	$0	$0	$0	$0	$0	$10,000	$15,000	$12,500
Metals	$16,240	$20,300	$18,270	$162,400	$203,000	$182,700	$162,250	$203,000	$177,625
HDPE Granulators	$8,010	$12,460	$10,235	$80,100	$124,600	$102,350	$267,000	$400,500	$333,750
Glass Crushers	$1,692	$4,230	$2,961	$16,920	$42,300	$29,610	$84,600	$211,500	$148,050
PROCESSING TOTAL	$42,690	$60,437	$51,563	$426,895	$604,365	$515,630	$1,016,275	$1,499,900	$1,258,088
ROLLING STOCK	$20,000	$25,000	$22,500	$200,000	$250,000	$225,000	$360,000	$500,000	$425,000
INSTALLATION	$19,669	$32,844	$26,256	$151,690	$202,749	$177,219	$261,302	$407,992	$334,647
CONTINGENCY	$23,636	$38,628	$31,132	$186,868	$298,711	$242,785	$387,758	$600,789	$494,273
TOTAL EQUIPMENT COST	$259,994	$424,908	$342,461	$2,055,443	$3,285,826	$2,670,634	$4,265,335	$6,608,681	$5,437,008
UNIT COST ($/TPD)	$25,999	$42,491	$34,245	$20,554	$32,858	$26,706	$8,531	$13,217	$10,874

2.5.3.3 Total Capital Costs

Estimated total capital costs by throughput capacity are presented in Table 2-31. The information in the table is divided into facility construction costs, equipment costs, and engineering fee in order to provide a range of total capital costs for each throughput capacity.
The ranges of total capital cost presented herein are at the upper end of the cost range for existing facilities. The reasons for this phenomenon are as follows:

- Many existing facilities do not have adequate floor area for unprocessed and processed material storage.
- Many existing facilities have been developed within existing structures, thereby avoiding stringent new building codes.
- The inclusion of commingled mixed paper in the facility designed for this manual increases capital costs for both sorting area and equipment. Most existing facilities do not have this capability.

2.5.4 Operating Costs

2.5.4.1 Labor Requirements

A range of labor requirements based upon facility throughput capacity is presented in Table 2-32. The data in the table show that the greatest variability is associated with the sorting function. Sorting efficiency (expressed as man-hours/ton) is highly dependent upon each particular facility design. In general, labor requirements for sorting per ton of material will decrease with increased capacity, due to the increased need for mechanical separation equipment such as classifiers and eddy current separators.
The number of sorters required also depends upon the degree of commingling of recyclable categories. A MRF which receives separated material categories (e.g., clear glass versus color mixed) will require significantly fewer sorters than those indicated in Table 2-32.

2.5.4.2 Operations and Maintenance

Operations and maintenance (O&M) costs are presented in Tables 2-33 and 2-34. Of the O&M cost elements listed in the tables, the costs that will vary the greatest include: 1) heating (which is a strong function of geographical location and degree of insulation); 2) maintenance (which is a function of type and quality of the equipment as well as diligence of routine maintenance); and 3) residue disposal.

Debt service has been included based upon an interest rate of 10% amortized over 20 years for facilities and seven years for equipment.

Taxes and depreciation have not been included in the tables due to their dependence on plant location and the tax structure of each particular business and financial arrangement. Consequently, the costs as presented may be considered appropriate for a publicly owned and operated MRF.

2.5.5 Sensitivity Of Capital And Operating Costs

As previously indicated, the capital and operating costs presented herein are based upon a recyclables stream which includes mixed waste paper (MWP) and old corrugated containers (OCC). The cost of sorting OCC and MWP from newspaper is substantial. If the list of recyclable materials is altered to eliminate MWP and OCC, the specifications for a given MRF capacity would be reduced as follows:

- Total floor area and construction capital costs would be reduced by 30%.
- Total sorting system costs would be reduced by 50%.
- Sorting labor effort and costs (including overhead) would be reduced by 50%.

The total impact on annual costs (including debt service) would be a reduction of over 30% when compared to costs included in Table 2-34.

2.6 PERFORMANCE GUARANTEES

Performance guarantees are established by the contractor, and normally become a part of the Agreement. The contractor is required t meet the guarantees presented throughout the course of the operating period.

2.6.1 Facility Availability

A contractor might guarantee, for example, that the MRF and its processing system would be capable of operating for 16 hrs per day for 6 days per week, if necessary. Any exceptions to this blanket guarantee should be incorporated in the Agreement.

Table 2-31. Estimated Equipment Cost by Throughput Capacity

Cost Item	10 TPD Low Cost ($/TPD)	10 TPD High Cost ($/TPD)	10 TPD Average Cost ($/TPD)	100 TPD Low Cost ($/TPD)	100 TPD High Cost ($/TPD)	100 TPD Average Cost ($/TPD)	500 TPD Low Cost ($/TPD)	500 TPD High Cost ($/TPD)	500 TPD Average Cost ($/TPD)
Construction	$29,563	$76,625	$61,150	$9,969	$29,494	$18,700	$6,325	$15,730	$10,780
Equipment	$25,999	$42,491	$34,245	$20,554	$32,858	$26,706	$8,531	$13,217	$10,874
Engineering	$6,667	$14,174	$10,247	$3,052	$6,235	$4,541	$1,188	$2,316	$1,732
TOTAL	$62,229	$132,290	$96,643	$33,575	$68,587	$49,947	$16,044	$31,263	$23,386

Cost Item	10 TPD Low Cost (Total $)	10 TPD High Cost (Total $)	10 TPD Average Cost (Total $)	100 TPD Low Cost (Total $)	100 TPD High Cost (Total $)	100 TPD Average Cost (Total $)	500 TPD Low Cost (Total $)	500 TPD High Cost (Total $)	500 TPD Average Cost (Total $)
Construction	$295,625	$766,250	$611,500	$996,876	$2,949,375	$1,870,000	$3,162,500	$7,865,000	$5,390,000
Equipment	$259,994	$424,908	$342,461	$2,055,443	$3,286,826	$2,670,634	$4,265,336	$6,608,681	$5,437,008
Engineering	$66,674	$141,739	$102,474	$305,232	$623,520	$454,063	$694,227	$1,167,894	$866,161
TOTAL	$622,294	$1,322,897	$956,426	$3,367,550	$6,868,721	$4,994,698	$8,022,061	$16,631,576	$11,693,169

Table 2-32. Typical MRF Labor Requirements.

| | Capacity (TPD) | | |
	10	100	500
Manager	1	1	1
Foreman/Operator	1	1-2	3-4
Sorters	1-2	13-25	60-80
Maintenance	0-1	1-2	4
Other a)	0	4-5	10-12
Administrative b)	0	1-2	2-3
TOTAL	3-6	21-37	80-104
Manhours/TPD Low	2.4	1.7	1.3
Manhours/TPD High	4.8	3.0	1.7
Manhours/TPD Average	3.6	2.3	1.5

a) Includes rolling stock operators, equipment monitors and cleanup staff.
b) Includes scale monitors, bookkeepers and clerical staff.

2.6.2 Facility Capacity

Guarantees on the capabilities of the processing system are required. An example of specifies coverage is as follows.

2.6.2.1 Paper Processing System

A guaranteed rated capacity, TPD, for newspaper, corrugated cardboard, office paper, and mixed paper would be established. Also, a guaranteed maximum process residue per ton of paper processed would be given. (Note: the process residue maximums cannot be practically guaranteed unless the collection system is properly managed.)

2.6.2.2 Commingled Processing System

A guaranteed rated capacity, TPD, for commingled material (paper excluded) would be established. Also, a guaranteed maximum process residue per ton of commingled material would be given.

Table 2-33. Typical MRF Unit Operating and Maintenance Costs.

Cost Item	Units	$/Unit
LABOR		
Sorters	(a)	$6.00/Hour
Other	(a)	$12.00/Hour
OVERHEAD b)	40% Labor	—
MAINTENANCE	—	$2.00/Ton (Low)
		$2.50/Ton (High)
INSURANCE c)	—	$3.00/Ton (Low)
		$4.00/Ton (High)
UTILITIES		
Power	15 KWH/Ton (Low)	0.04 $/KWH (Low)
	20 KWH/Ton (High)	0.07 $/KWH (High)
Water & Sewage	70 GPD/Person	$2.00/1000 Gal
Heating d)	0 MBTU/Ton (Low)	$4.00 $/MBTU (Low)
	0.05 MBTU/Ton (High)	$8.00 $/MBTU (High)
FUEL	0.2 Gal/Ton	$1.20/Gal
OUTSIDE SERVICES & SUPPLIES	10% Operating Costs	—
RESIDUE DISPOSAL	0.1 Ton/Ton	$25.00 $/Ton (Low)
		$100.00 $/Ton (High)

a) Varies based on number of employees per Ton. See Table 2-30.
b) Includes Social Security, vacation and sick leave and insurance.
c) Includes workers' compensation, property and liability.
d) Range of use based on climatic extremes.

Table 2-34. Estimated Annual O&M Costs by Throughput Capacity

Equipment Item	10 TPD Low	High	Average	100 TPD Low	High	Average	500 TPD Low Cost	High Cost	Average Cost
LABOR									
Sorters	$12,480	$24,960	$18,720	$182,240	$312,000	$237,120	$748,800	$998,400	$873,600
Other	$49,920	$99,840	$74,880	$199,680	$299,520	$249,600	$499,200	$599,040	$649,120
OVERHEAD a)	$24,960	$49,920	$37,440	$144,768	$244,608	$194,688	$499,200	$638,976	$569,088
MAINTENANCE	$5,200	$8,500	$5,850	$52,000	$65,000	$58,500	$260,000	$325,000	$292,500
INSURANCE b)	$7,800	$10,400	$9,100	$78,000	$104,000	$91,000	$300,000	$520,000	$455,000
UTILITIES									
Power	$1,580	$3,840	$2,600	$15,800	$36,400	$26,000	$78,000	$182,000	$130,000
Water & Sewage	$36	$73	$55	$473	$910	$692	$2,184	$2,912	$2,548
Heating c)	$0	$1,402	$701	$0	$14,018	$7,008	$0	$70,080	$35,040
FUEL	$624	$624	$624	$8,240	$8,240	$8,240	$31,200	$31,200	$31,200
OUTSIDE SERVICES & SUPPLIES	$10,268	$19,738	$14,997	$65,900	$108,269	$87,085	$260,868	$336,761	$293,810
O & M SUBTOTAL	$112,838	$217,094	$164,066	$724,901	$1,190,863	$957,932	$2,759,442	$3,704,369	$3,231,908
O & M COST ($/TPD)	$43.40	$83.50	$63.45	$27.88	$45.81	$38.84	$21.23	$28.50	$24.88
RESIDUE DISPOSAL	$6,500	$26,000	$16,250	$65,000	$280,000	$182,500	$325,000	$1,300,000	$812,500
DEBT SERVICE	$93,749	$188,635	$139,319	$660,258	$1,068,325	$801,155	$1,284,745	$2,381,396	$2,180,344
TOTAL ANNUAL COST	$213,087	$431,729	$320,538	$1,350,150	$2,519,208	$1,921,588	$4,369,187	$7,365,764	$6,224,749
ANNUAL COST ($/TPD)	$81.96	$166.05	$123.28	$51.93	$96.90	$73.91	$33.61	$56.66	$47.88

a) Includes Social Security, vacation and sick leave and insurance.
b) Includes workers' compensation, property and liability.
c) Range of use based on climatic extremes.

Guarantees would be made that material specifications would be met. These specifications would include glass, aluminum, ferrous and plastics; specifications would be detailed in the Agreement.

2.6.3 Environmental Guarantee

The contractor would guarantee that all components of the facility would comply with all applicable federal and state ordinances, rules and regulations, and any federal, state, or county permits, licenses, or approvals issued with respect to the facility.

2.7 MARKETING

It is readily apparent to anyone in the recycling field that stable markets for collected materials are vital to any successful program. The recycling movement has increased in popularity throughout the United States; however, it has brought with it a need to ensure that once materials are collected there will be a market for them. Skeptics of any waste management practices utilizing recycling will habitually ask the questions, "what happens if you lose your market?" (In fact, this is a very important questions; established secure markets are vital to any successful MRF operation.) this attitude also prevails in the political arena where states or municipalities who are committed to recycling major portions of the solid waste stream are asking these same questions regarding the disposition of the recycled material.

An encouraging factor for the future of recycling is the rapid rise in waste disposal costs over the last few years throughout the country. This "avoided cost" situation has favorably altered the economics of recycling; however, much of the industry will survive only if the revenue from the sale of recyclable materials is sufficient. (It should be noted that there is some indication of declining costs for disposal in some areas of the country due to source reduction, recycling, recession, etc. diverting waste from landfills and incinerators, and creating "shortages of waste.")

2.7.1 Market Concerns for Recovered Wastepaper

At the present time, the waste paper market is one of the recycling industry's primary concerns. Over the past year or two the industry has blamed successful residential curbside collection programs for causing a glut in the market place, and a subsequent recession in the waste paper markets.

The waste paper markets have experiences dramatic downturns in the past; these downturns occurred in the early 70s and again in the early 80s. Therefore, the industry has experiences ups and downs long before residential collection programs ere ever instituted. (Government supported residential collection programs were essentially nonexistent in the 70s and early 80s.) yet today some industry representatives are blaming local governments for the potential demise of the waste paper business. However, most large city governments have no choice but to develop aggressive recycling programs as a means of reducing operational costs, extend landfill life, and reduce the environmental hazards of landfilling.

Newsprint manufacturers are now receiving pressure from their customers to use more and more recycled newsprint; this pressure may result in additional production of newsprint containing recycled fiber. In another paper arena, capacity is growing for the use of old corrugated cardboard (OCC). The utilization of OCC at the manufacturing mills now exceeds a million tons per year. The trend for OCC use is positive; it has grown at a rate of 12 percent per year over the past several years. In addition recycling of high-grade office paper has grown at a rate of 4 1/2 percent per year over the last 10 years. Mixed paper markets show less promise; also, decreases in packaging (source reduction) may take away some mixed paper markets. A compilation of waste recovery figures for 1989 and 1990 is shown in Table 2-35.

It does appear that markets for recyclable paper products are adjusting to this supply increase that will allow more and more citizens to participate in waste reduction. Markets in the United States in the past have responded to the public demand for consumer products, and hopefully will respond to the public desire to reduce waste and purchase recycled products.

2.7.2 Market Concerns for Recovered Steel Containers

A ready market exists for steel cans. When discussing the recycling of steel cans, reference is made to two types of cans: (1) the common tin can (tin plated) that is widely used for foodstuffs, etc.; and (2) the bimetallic can (steel can with aluminum top) that is used for carbonated beverages. A most important point to remember is that steel scrap has been an essential ingredient in steel making for some time. In fact, the process is designed to utilize steel scrap, so that the market for steel cans should continue to be dependable and very likely an expanding one.

As is found with other secondary commodity materials, steel can prices will vary according to market demand and geographic region. Because of the world wide market demand, prices for established grades of iron and steel scrap are published regularly in a number of national publications. For example, steel can prices for baled railcar quantities are published in the scrap iron and steel prices section of American Metal Market and in Iron Age.

Table 2-35. Waste Paper Recovery Figures.

1990 Waste Paper Data
(000 Short Tons)

	Consumption at U.S. Paper and Paperboard Mills*	Exports	Total Collected
News	4,679.2	1,256.7	5,935.9
Corrugated	10,447.7	2,730.8	13,178.5
Mixed	2,491.9	1,146.3	3,638.2
High Grades	4,761.7	1,371.1	6,132.8
TOTAL	22,380.5	6,504.9	28,885.4

1989 Waste Paper Data
(000 Short Tons)

	Consumption at U.S. Paper and Paperboard Mills*	Exports	Total Collected
News	4,138.1	1,281.1	5,419.2
Corrugated	9,993.5	2,918.8	12,912.3
Mixed	2,355.6	853.7	3,209.3
High Grades	4,455.1	1,253.4	5,708.5
TOTAL	20,942.3	6,307.0	27,249.3

* Includes consumption of molded pulp and other nonpaper uses.
   (American Paper Institute)

Markets for all recycled materials including steel cans are essentially regional in nature. The Steel Can Recycling Institute (SCRI) maintains an up-to-date list of known purchasers throughout the country for steel. This information base is constantly expanded as new community programs come on-line. It is important then to contact SCRI directly to get the most current information on scrap steel prices.

Generally speaking, the buyers in closest geographic proximity to a community will be the most logical purchasers of steel cans. An exception to this general rule is the large national detinning companies which have their own transportation networks, and are presently working to establish regional buying networks for steel cans. Establishment of this highly cost effective transportation system allows communities to market their steel cans to plants that are hundreds of miles away from them.

Steel mills are prime marketers for steel cans, but there are other big potential markets including detinning companies. These detinning companies have been working for a long time in recycling tin cans. (Direct purchases by steel mills are impacting detinning economics.) Iron and steel foundries are also part of the nation's steel-making infrastructure. They have not historically used a lot of steel cans, but the forecast indicates that this type of market for recovering steel cans will be an active one in the years ahead. Scrap processors and dealers are other potential markets for steel cans. They have been supplying the industry with scrap material for many years, and their role on scrap recycling and utilizing the cans is one that looks as if it will increase in the future.

2.7.2.1 Steel Mills as Ultimate Market for Steel Cans

Steel mills are the major users for most steel cans; there are more than 120 steel mills in the United States that have operating furnaces. The steel-making process allows a certain amount of tin in the scrap mix; also, mills can combine steel can scrap with other scrap sources to produce new steel. The steel industry has been recycling scrap steel heavily through the 80s; in fact, approximately 100 billion pounds of used steel were remelted each year in the 80s.

Steel mills have essentially two types of furnaces: (1) the basic oxygen furnace which utilizes approximately 20-30 percent recycled steel scrap, and (2) the electric arc furnace which uses nearly 100 percent scrap. As time goes on, steel cans are becoming more and more an essential part of the scrap mix.

In preparing steel cans for market, the method used will depend pretty much on the end market. For example, cans can be shipped loose, shredded, or baled loosely or densely. Also, the end markets do not necessarily need to receive cans with labels and ends removed; and steel mills are generally tolerant of small levels of foreign matter. Paper labels and small amounts of plastic found on the tops of aerosol containers, for instance, are burned in the extremely high temperatures of the furnace, so there is really no need for concern for contamination from this material. Bimetal cans, also do not require any special preparation for sale to steel mills. They should be collected and processed mixed with all other types of steel cans; in fact, the aluminum found on the tops of steel beverage cans actually enhance the steel making process.

2.7.2.2 Detinning Companies as Ultimate Market for Steel Cans

In addition to the steel mills, detinners also purchase steel cans directly. Most of them have sophisticated equipment that shreds the cans so that paper labels and other minor contaminants are removed prior to detinning. Through various processes, detinners remove the tin from steel products containing steel. Then they sell the detinned steel to steel mills and foundries, and the recovered tin to its appropriate markets. Each steel can purchaser whether it be a steel company, a foundry, a detinning company, or whatever, has its own specifications for postconsumer steel cans. In each category, the steel can scrap may include aluminum lids, but generally excludes nonmetallics or other nonferrous metals, except those used in can construction.

2.7.3 Market Concerns for Recovered Glass

There are a number of ways glass bottles can be reused. They can be ingredients in the making of fiber glass and reflective beading; they have also been used to help control beach front erosion and as a substitute for stone in the making of roadway "glasphalt." However, the most logical market for used glass containers is a glass plant similar to the one where they were manufactured. At a glass plant they can be melted down and remade as new bottles and jars in a true example of closed-loop recycling. Nearly all plants purchase glass from the general public; therefore, for any beginning recycling project a glass plant is the ideal spot to sell bottles and jars. For those who are not near a glass plant, a call to one of the many intermediate glass brokers would be in order.

When contacting the plant or broker, it is advisable to determine the hours of operation, prices paid, and any particular quality requirements. Most plants will provide a specification sheet upon request.

If a recycler has substantial tonnage of cullet (broken glass) to sell, he may be referred to the company recycling director to make special arrangements.   An investigation of the market will show that glass recycling specifications are rather straightforward. It is most important that the material be color sorted and contaminant free. The question then might arise, what is color sorted? For example, would a load of green glass be rejected if it has one amber container? Also, just what are the contaminants that are of concern.

Color sorting is truly essential to the operation of a glass plant; because it is most important to assure that newly manufactured containers match the color specifications required by the customer. For example, if too much amber glass is put into a clear flint batch, it can result in off-color bottles. Further, mixed color cullet can cause chemical composition problems; it can interfere with the redness ratio which controls light transmission through a container. With large amounts of contaminants, reactions between the reducing and oxidizing agents found in brown and green glass can create foaming in a melting furnace. Nevertheless, some markets do exist for mixed color cullet, especially in the fiber glass industry. However, those markets are neither as stable nor as lucrative as those for color sorted glass. Occasionally one hears talk of an "ecology" bottle; it is made entirely from mixed color cullet, but such a container finds few buyers in today's market place.

There is some tolerance in color separation; and specifications will vary from plant to plant.   However, in general to process glass into furnace-ready cullet so that it can be used directly in the manufacture of new glass containers (bear in mind that these guidelines are not necessarily acceptable for all consumers

- only container glass is acceptable;
- glass must be separated by color into flint (clear), amber (brown), and green;
- in flint glass, only 5 percent of the total load can be colors other than flint; in amber glass 10 percent; and in green glass up to 20 percent;
- glass must be free of any refractory materials; it will be rejected if there is more than a trivial amount of ceramic material; and

- glass must be free of metallic fragments and objects, dirt, excessive amounts of paper, or large amounts of excessively decorated glass.

As previously stated, there is an excellent market for contaminant free cullet; however, practically no market exists for contaminated cullet. Some of the contaminants that most effect a glass plant operation are metal caps and lids, ceramics, stones, and dirt. In the making of new glass containers, silica sand, soda ash, and limestone are the primary raw ingredients. Cullet can be added to this mixture, which is then heated to approximately 2600°F. At this temperature the batch mixture is turned into a fiery molten state that can be formed into bottles and jars; however, metals, ceramics, and stones do not melt. Instead, they remain intact and can damage the glass melting furnace, or appear in the new containers that are being made. Ceramics are especially bad because they may breakup into countless fragments; and, they are not usually found until they show up in the newly made containers. These imperfections would normally be caught by inspection before they leave the plant; however, at this point they have already created a major manufacturing problem.

Another source of contamination is the ceramic and wire caps that are found on some beer and wine bottles. Since the caps remain attached to the neck of the bottle, they often end up in the recycling bin, and, subsequently at the glass plant. Most of the nation's glass plants have beneficiation units on site, or nearby, that will remove metal contaminants, as well as plastic and paper labels. However, these units won't detect ceramics or stones. Thus, a solution to this potential problem must depend on careful processing by the supplier of the cullet. Although paper and plastic labels do not need to be removed, the bottles should be lightly rinsed. As a rule of thumb, a bottle clean enough to be stored in a home while awaiting collection should be clean enough to be recycled. On another note, it is Well to be aware that container glass is what the glass plants need and want. Heat resistant glass, along with windshields, windows, and crystal, should never be mixed with bottles and jars as their ingredients are different, and the glass plants do not want them.

Glass plants do not require that glass bottles and jars be crushed by recyclers. The main reason for crushing by a recycler is to minimize volume for ease in handling and transportation. Therefore, the question might arise as to whether a beginning recycler should invest in a glass crusher. The answer would depend on factors such as the volume expected, the type of transportation to be utilized, and the distance to the market. Usually if glass is to be shipped as bulk in dump trucks, it is not necessary to crush the glass. Many recyclers ship glass in "gaylord" boxes; these boxes can hold as much as a ton of cullet, and are designed for a fork lift operation. However, for high volume users, it is usually advisable to ship by dump truck, or even by rail.

The future of the glass cullet market looks promising. At present, usage is estimated to be 25 to 30 percent; however, the industry has announced an overall goal of 50 percent cullet usage, and has setup glass recycling programs over much of the United States. Many plants could increase their cullet consumption to the 70 to 75 percent range, if there were adequate supplies of cullet. For the foreseeable future, cullet prices should remain relatively stable. The price of cullet reflects the avoided cost of raw materials, and the energy savings for the lower melting temperature of used glass. Unlike aluminum, glass cullet is not actively traded on the world markets; so, does not fluctuate due to international demand or currency rates. Perhaps the biggest question is, can recyclers provide the quantity of quality color sorted glass that the plants need? Curbside programs that commingle material are bound to produce some glass residue that can not be color sorted. Glass container plants will be unable to accept this residue and alternative markets must be pursued.

2.7.4 Market Concerns for Recovered Aluminum

The recycling in the United States of aluminum used beverage cans (UBCs) continues to increase. This trend is shown in Table 2-36, where the 1990 recycling rate is listed as 63.6 percent. Further, this translates to a recycling of 54.9 billion cans, with the recovery of some 1.93 billion pounds of aluminum. As shown in Table 2.36, aluminum UBC recycling has been increasing dramatically for several years, especially during the 1980s.

Most of this recovered aluminum has gone directly back into new cans, because it is possible to make an aluminum can entirely of recycled metal. Typically, an aluminum can body is made from used aluminum beverage cans and can manufacturing scrap. However, primary aluminum (from the ore) is needed as the total volume demand considerably exceeds the supply of recycled metal. Aluminum can ends are typically made from alloyed primary aluminum and end manufacturing scrap. Therefore, it is possible to have a finished aluminum can and end that come almost entirely from recycled sources. It is likely, however, that the newly manufactured aluminum can will be produced from a mixture of recycled aluminum, can manufacturing scrap and primary aluminum in percentages dictated by company needs, production schedules and market economics at the time.

A point to remember, however, is that every aluminum can that is recycled can go into a new can. This situation assures a never ending market for a container that does have the best recycling record in the beverage industry. Further, aluminum produced by recycling requires 95 percent less energy than that needed to make it from the ore. This contributes to a scrap value that makes recycling possible without any kind of corporate subsidy or government assistance. No other beverage container material has the capability, as does aluminum, to pay the public a sufficient amount of money to motivate them to recycle. The value is there, the market is there; aluminum can recycling will work.

Table 2-36. U.S. Aluminum UBC Recycling Rates.

Year	Million Lbs.	Billion Cans	Recycling %
1972	53	1.2	15.4
1980	609	14.8	37.3
1985	1,245	33.1	51.0
1989	1,688	49.4	60.8
1990	1,934	54.9	63.6

Calculation for 1991 rate:

UBC scrap (billion lbs.)	1.934
Average number of cans/lb.	28.43
Total cans recycled (billions)	54.984
Total new cans shipped (billions)	86.513
Recycling rate	63.6

(Aluminum Assn., Can Manufacturers Institute, Institute of Scrap Recycling Industries)

All major beer brands and most soft drinks are sold in aluminum cans; about 95 percent of today's beverage cans are aluminum. In addition, most cans are clearly labeled as recyclable aluminum. However, this can be verified by placing a magnet on the side of the can (will not stick to aluminum). Aluminum cans must be clean and dry for recycling, or most recycling centers will deduct 10 percent from the purchase price for dirty or wet containers. Further, it is well to remember to keep the collected cans in a secure place, indoors if possible. Used cans are valuable, and should not be transported in a vehicle open to wind and weather.

CHAPTER 3

General MRF Concerns

3.1 SITING AND PERMITTED CONCERNS

Several criteria need to be considered when locating a MRF. First of all, it is desirable that the MRF be near the collection area, since minimization of travel distances is quite important to the successful operation of a MRF. In addition to proximity to the collection routes, access to major haul routes is also important. Access roads must be able to handle heavy truck traffic; also, truck routes should be designed to minimize the impact of vehicular traffic on surrounding neighborhoods. Aside from the routing issues, the land on which the MRF is to be built must be zoned for industrial purposes, and the area used should provide satisfactory isolation. Further, it is most important when siting a facility to involve the neighborhood, and secure community acceptance. This is, in many cases, the most difficult task in the siting procedure. Of interest is the fact that some communities have had good success in using closed landfill sites as sites for new MRFs.

In the past, the decision-making process for situations concerning municipal solid waste management was normally centralized in the hands of a few key governmental personnel. However, over the last 20 years or so, nongovernmental interests have become more involved in local decision-making; and, citizens have demonstrated that they will not accept "behind the scenes" decisions on solid waste management. Therefore, the manner in which the siting process is carried out for a MRF can have a significant effect on public acceptance of the overall project by the public. Not only can a closed-door, decision-making process waste time and resources, it can jeopardize the credibility of the professional planners. Further, if the trust and confidence of the public is lost, it is nearly impossible to recover.

The siting process normally consists of three related phases: planning, site selection and facility design, and implementation. Any of these stages of the siting process may be subjected to intense public comment and debate. A review of the major steps in facility siting (EPA/530-SW-90-019) show that important decisions are made very early in the planning phase for a solid waste management facility (Figure 3-1).

For example, early in the planning phase, choices must be made to determine a waste management strategy, and whether a MRF is the facility that is really needed. Later, after a decision has been made to site a MRF, a major issue remains as to its location. The criteria that are assessed to determine suitability for a potential site include hydrogeological conditions, socioeconomic characteristics, and population densities. Regardless of where the MRF is located, the burden of the facility will be placed on the people living nearby; thus, exposing them to more noise, traffic, and pollution than the overall population being served by the facility. Sometimes these constituencies are rural or impoverished people who tend to be poorly represented in the traditional decision-making process. Nevertheless, these people can gain the support of a large coalition from within or outside the community in response to potential inequities or other political issues.

In selecting a site for a MRF, some citizens will almost certainly question the validity of any technical work carried out. Also, the involved community will be concerned about negative effects on property values, safety, air quality, noise, and litter; or about broader issues such as the impact on community prestige. Some citizens may argue for compensation arrangements, or other forms of guarantees against negative impacts. Also, it is normal for public opposition to increase as site selection time approaches. Finally, before a site is selected, the overall project must be approved by state agencies that are often responsive to political pressure from community groups.

The public concerns are usually associated with safety features of the facility. Groundwater contamination and air pollution are by far the issues most frequently requiring attention, although noise, litter, and traffic issues also appear. The operator's credentials and past record are also important concerns during site selection or facility design. Other points of contention may include the types of wastes allowed at the site, and whether the site should be restricted to local haulers.

Phase I: Planning

• Identifying the Problem	Recognizing the growing waste stream, rising costs, and capacity shortfall
•Designing the Siting Strategy	Planning and integrating public involvement, risk communication, mitigation, and evaluation activities.
•Assessing Alternatives	Researching, debating, and choosing among the options: recycling, source reduction, incineration, andland disposal.
•Choosing Site Feasibility Criteria	Studying population densities, hydrogeological conditions, and socioeconomic characteristics.

Phase II: Site Selection and Facility Design

•Selecting the Site	Performing initial site screening and designation; acquiring land; conductiong permit procedures; developing environmental impact statements.
•Designing the Facility	Choosing technologies, dimensions, safety characteristics, restrictions, mitigation plans, compensation arrangements, and construction.

Phase III: Implementation

•Operation	Monitoring incoming waste; managing waste disposal; performing visual and lab testing; controlling noise, litter, and odor.
•Management	Monitoring operations and safety features; performing random testing of waste; enforcing permit conditions.
•Closure and Future Land Uses	Closing and securing the facility; deciding on future land uses; and performing continued monitoring

Figure 3-1. The three-phase siting framework

Operation and management plans for a MRF often are important to the general public. Demands are sometimes made for strict monitoring and enforcement activities to ensure compliance by haulers and operators. These demands may include local supervision of the facility, along with state agencies' support of the local enforcement efforts. These actions may include revoking disposal permits, testing wastes, and monitoring air and groundwater. It is also important to note that no siting proposal is complete without planning for closure and future land use. Local citizenry will often argue how the land should be used after closure, or how groundwater monitoring should be maintained.

Issues and challenges facing public officials and citizens have changed over the last two decades. It is reasonable to expect that new issues and new challenges will emerge in the coming years. There is no set of procedural steps that will guarantee a successful siting process. Public officials from different communities must tailer their siting strategy to their own particular needs and issues. The following guidelines summarize the most important points made in this discussion:

- accept the public as a legitimate partner;
- listen to the concerns of the different interests and groups in the community;
- plan a siting process that permits full consideration of policy alternatives;
- set goals and objectives for public communication activities in each step of the process;
- create mechanisms for involving the public early in decision-making process;
- provide risk information that the public needs to make informed decisions;
- be prepared to mitigate negative impacts on the community; and
- evaluate the effectiveness of public involvement and risk communication activities.

Although these eight guidelines are not all-encompassing, each is important in defining an effective siting process. The guidelines are specific enough to lend structure to a multitude of planning activities, but they do not substitute for the good judgement of project leaders and other interested parties.

3.2 CONTRACTING ISSUES

Unlike air and water pollution control which has been largely regulated at the state and federal levels, solid waste disposal has traditionally been the responsibility of local governments (although now regulated, to a degree, at both state and federal levels). However, the design, construction and operation of a MRF is more like a general business enterprise than are the more traditional municipal functions, such as public health and safety, social services, etc. Nevertheless, there are now a growing number of private/public partnerships in the MRF industry that illustrate the utilization of the resources and capabilities of a public agency, while enjoying the greater flexibility and efficiency associated with private sector operations.

When a local government undertakes establishing a MRF as a means of reducing the solid waste disposal stream, it must first assess its own capabilities and then define the role of any prospective private partner. Conversely, however, it may be in the interest of a private developer to attempt to interest a public agency in such an endeavor. In any event, promoters of any kind of recycling initiatives often have to abide by public procurement and contracting procedures that have been dictated by state and local law.

3.2.1 Contractual Arrangements

Before entering into any type of MRF contractual arrangement, the sponsor of such a program must address certain issues: (1) what recyclable materials are actually present in the waste stream, and their quantities; (2) is the processing facility going to be directed toward centralized mechanical processing or more toward source separation; and (3) what is the relationship with the markets, and the quantity and quality of recovered materials that can be sold.

A successful recycling project will often involve some form of joint venture between a public agency and a private contractor. Among the MRFs which are currently operational, approximately 65 percent are owned by private firms with the remaining facilities owned by the public or not-for-profit sectors (GAA, 1990). However, with regard to the planned facilities, the ownership picture changes substantially; 62 percent of the planned facilities will be publicly owned, with the private sector decreasing its share to 38 percent of the projects. However, despite this trend toward public ownership and financing, private firms will continue to operate most of these facilities. Private enterprises operate 83 percent of the existing projects, and about the same (79 percent) of planned installations (GAA, 1990). A more recent survey (Biocycle, 1991) showed a reversal in this trend with about 73 percent of all operating facilities being privately owned, with 82 percent being privately operated.

A formal procurement aimed at establishing a MRF can involve: (1) either a two-step process, where responses to a request for qualifications (RFQ) are evaluated to establish a short list of qualified contractors who are eligible to respond to a request for proposal (RFP), or (2) a combined RFQ/RFP under which each firm making a proposal has to establish its qualifications in the course of extending its offer. Whether contractors are screened first in an RFQ or as part of an RFP, the importance of selecting a qualified party can not be overemphasized. There is no substitute for a contractor having the appropriate skills, experience, and technical and financial resources to implement a project effectively. In order to give an idea of the scope of an RFP, a sample Table of Contents for an RFP is shown in Table 3.1.

The RFP should present as much background information as possible concerning the project; any contractor before submitting a response to an RFP will want to know that the proposed facility has a good chance of being financed and built. The background section of the RFP should include a discussion of a number of topics. For example, the RFP should address the demographic and economic characteristics of the area, legal authority of the procuring agency, the type of guarantee with regard to the waste supply, information concerning any private recycling programs that are taking place in this area, and what sort of public support and regulatory requirements are to be expected.

Next, the RFP should detail the respective responsibilities of the procuring agency and the contractor. Normally the municipality would be involved in furnishing the facility site, and providing information necessary for securing any types of environmental permits for construction and operation of the facility. The municipality would also usually be responsible for delivering, or having delivered, the recyclable material to the facility. The municipality probably would also be in a position of exercising legal control over the disposal of waste materials from the MRF. Further, a typical RFP will normally assign to the contractor any risks and responsibilities involved in developing the project. These responsibilities can include design and construction, and furnishing of the labor, supplies, materials, equipment, services, and technology necessary to complete the facility in accordance with product specifications.

Table 3-1. Sample Table of Contents for an RFP (To Receive, Process, and Market
Household Recyclable Materials)

CONTENTS

1 General Information

    1.1   Introduction
    1.2   Plan Implementation
    1.3   Overall Program Timing

2 Project Overview

    2.1   Introduction
    2.2   Recycling Implementation Plan
    2.3   Procurement of MRF Services
    2.4   Recyclable Materials Collection and Delivery
    2.5   Recyclable Materials Quality
    2.6   Recyclable Materials Quantities
    2.7   General Requirements

3 Technical Requirements

    3.1   Facility Requirements
    3.2   Operations Requirements
    3.3   Environmental Performance Standards
    3.4   MRF Public Education Facility Requirements
    3.5   Proposer Technical Experience and Qualifications

4 Service Requirements and Business Arrangements

    4.1   General
    4.2   Service Requirements for Facility Siting, Permitting Design, and Construction
    4.3   Service Requirements for Facility Operations
    4.4   Option to Provide Services to Private Customers
    4.5   Term of Service
    4.6   Performance Guarantees and Assurances
    4.7   Financing
    4.8   Performance Bonds and Proposal Security
    4.9   Payment for Services
    4.10 Business Proposal
    4.11 Default and Remedies
    4.12 Insurance Requirements
    4.13 Other Requirements
    4.14 Minimum Financial Qualifications
    4.15 City Policy Compliance

5 Proposal Requirements and Evaluation

    5.1   Executive Summary
    5.2   Qualifications of Proposer and Project Organization
    5.3   Technical Proposal

Table 3-1. Sample Table of Contents for an RFP (To Receive, Process, and Market Household Recyclable Materials) (CONTINUED)

5.4   Business Proposal
5.5   City Policy Compliance
5.6   Proposal Evaluation Criteria

APPENDICES

Appendix A Historical Daily Tonnages (FY 1987-88)
Appendix B Technical and Business Proposal Forms
Appendix C City Policy and Compliance Attachments

The RFP should contain a section dealing with the criteria under which the proposals will be evaluated. Points that are normally covered in the criteria include the following: (1) technical feasibility of the facility design; (2) prior experience with this design, and whether or not similar facilities have been operated elsewhere; (3) qualifications of the personnel assigned to the project; (4) efficiency and reliability of the proposed system, with special attention to the subjects of safety and environmental protection; (5) credit rating and financial stability of the proposing party; and (6) net revenue or net cost that would be imposed on the procuring agency.

Although the RFP and the resultant proposal tend to be lengthy, complex documents, the end result is an offer by the proposer to the public agency to perform certain work for a specified price under terms and conditions established in the RFP. (A sample proposal Table of Contents is shown in Table 3.2.) Further, it is certain that an effective public/private sector partnership depends on a clear understanding by each party of its respective rights and obligations.

3.2.2 Flow Control

Municipalities are now being forced to resort to waste disposal methods other than landfilling. This has come about largely due to the shortage of landfill capacity in the United States. Recycling, composting, and other types of approaches to municipal waste stream management are being explored extensively. When considering a MRF, it is essential that the municipality be able to guarantee delivery of consistent amounts of solid waste. This can be accomplished by a municipality only if it can control the waste streams within its boundaries. It is normal practice,

Table 3-2. Sample Table of Contents for a Proposal.

Sample Elements of a Proposal

Proposer Qualifications, Technical Proposal
Business Proposal, City Policy Compliance

CONTENTS

Section I--Proposer Qualifications

1.0   Introduction

1.1   Project Team Experience

    1.1.1 Design and Technical Qualifications
    1.1.2 Reference Facility

1.2   Project Team Organization

    1.2.1 Organizational Chart
    1.2.2 Design/Equip Team
    1.2.3 Operations Management
    1.2.4 Project Team Staffing

1.3   Local Employment Opportunities, Local Business Involvement

1.4   Financial Qualifications

1.5   Personnel and Facility Management

1.6   Marketing Management

1.7   Technical Ability

Section II--Technical Proposal

2.0   Introduction

2.1   Location

2.2   General Design

    2.2.1 Building Description

        2.2.1.1 Architecture
        2.2.1.2 Building Description
        2.2.1.3 Public Education Facility Description
        2.2.1.4 Building Structures, Utilities, and Details

Table 3-2. Sample Table of Contents for a Proposal. (CONTINUED)

Table 3-2. Sample Table of Contents for a Proposal. (CONTINUED)

2.13 Process Control and Instrumentation

2.14 Process Mass Balance

2.15 Process Energy and Water Balance

2.16 Availability Analysis

    2.16.1 System Availability
    2.16.2 Rugged Engineering
    2.16.3 Built-in Surge Capacity
    2.16.4 Contingency Sorting and Processing Strategies

2.17 System Capacity

    2.17.1 Expanding Minimum Design Capacity--Mixed Recyclables
    2.17.2 Expanding Minimum Design Capacity--Paper Line
    2.17.3 Expanding Minimum Design Capacity--2 Shift Operation

2.18 Product Specification:

    Glass, Aluminum Tin, PET, HDPE

2.19 Materials Marketing

    Aluminum, PET, HDPE, Tin, Glass, Mixed Cullet, Newspaper, Letters of Intent

Facility Drawings and Schedules

    Site Plan Layout
    General Arrangement
    Facility Cross Section
    Electrical Single-Line Diagram
    Schedule

Technical Proposal Forms

    B-2   Technical Description of Site/Facility/Equipment
    B-11   Performance Guarantees
    B-12   Performance Assurances
    B-13   Product Specifications

Section III--Business Proposal

    Proposal Forms

Table 3-2. Sample Table of Contents for a Proposal. (CONTINUED)

Section IV--City Policy Compliance

4.0   Contractor's Past Record

4.1   Construction Phase Compliance

4.2   Local Employment Opportunities

4.3   Operations Phase Compliance

4.4 Non-Profit Organization Involvement

Appendices

Appendix A: Resumes of Key Project Team Members
Appendix B: Throughput Verification and Material Storage Calculations

List of Tables

Table 1: Contractor'"s Recycling Facilities
Table 2: Project Organization Chart
Table 3: Operations Staffing Chart
Table 4: Facility Throughput Capacities
Table 5: Processed Material Loadout Chart
Table 6: Facility Staffing
Table 7: Potential Future Recyclable Materials
Table 8: Revenue Projections

however, that a municipality in controlling the MSW stream within its boundaries controls both the disposal site of the waste and the price paid for disposal.

Municipalities typically exercise control over waste flow by passing legislation that requires haulers to transport the solid waste they have collected to a disposal site that is determined by the municipality. The haulers pay a tipping fee at the disposal site; the fee is established by the municipality. However, competitive concerns may be raised because this legislation, passed by the municipality, might allow the municipality to essentially control the entire market for solid waste disposal. Accordingly, activities of this type could come within the jurisdiction of the federal antitrust laws. State and local governments, simply because they are governmental entities, are not automatically exempt from federal antitrust laws. In addition, private parties that contract with such governmental entities for waste disposal services are also potentially liable under federal antitrust laws.

Now while state governments may be exempt, municipal governments do not necessarily receive a blanket exemption because they have only delegated, not sovereign power. However, the United States Supreme Court has ruled that municipalities qualify for the state action exemption so long as their anticompetitive behavior is undertaken pursuant to a clearly articulated state policy. Further, protection from antitrust damage liability is also available under the Local Government Antitrust Act of 1984, which prohibits damage actions against local government officials and employees acting in an official capacity as well as private persons acting at government direction.

Once it has been determined that it is legal to control the waste stream, the next step is to determine how this control is to be exercised. A municipality does have a variety of options with regard to controlling the waste stream. First, control can be exercised merely by requiring its drivers to haul the waste to a specific site. Second, the municipality might have a contract with the haulers which would authorize them to haul the waste to a specific site. Or thirdly, the municipality could authorize private collection by allowing a direct contractual arrangement between the residents and the private haulers.

Another area where the municipality must retain control is the cost of disposal. The municipality would typically develop an annual rate setting procedure based on estimated cost of the disposal system and estimated revenues from the system. This rate determination procedure would normally be administered by an agency that has been given the authority to manage the solid waste system.

Waste flow control to a MRF is necessary not only for financial reasons, but in order for the system to operate efficiently. A MRF would be developed to handle a specific level of throughput. If that level of waste is not available, costly downtime results for the facility. An insufficient amount of waste also would increase the cost per ton of waste handled, and would have a detrimental effect on equipment maintenance schedules. It is clear that municipal control over the waste stream is essential to an efficient waste disposal system and the lack of control can lead to unintended consequences. If waste flow control and proper administration is carried out, then the municipality will better be able to ensure that its waste disposal system can operate as anticipated.

APPENDIX A

Glossary

(Definitions drawn prinicpally from ASTM Special Technical Publication 832, H.I. Hollander, ed.)

Acceptance testing: Testing of process equipment and the overall processing system.

Aggregate: A granular material of mineral composition such as sand, gravel, shell, slag, or crushed stone used with a cementing medium to form mortars or concrete, or alone as in base courses, railroad ballasts, etc.

Air classification: A process in which a stream of air is used to separate mixed material according to the size, density, and aerodynamic drag of the pieces.

Air classifier: A mechanical device using air currents to separate solid components into "light-fraction" or "heavy-fraction."

Air dry: Paper or paperboard is air dry when its moisture content is in equilibrium with atmospheric conditions to which it is exposed. According to trade custom air dry pulps are assumed to contain 10% moisture, and are sold on this basis.

Air knife: Jargon for a blower device intended to separate steel cans from more massive pieces or iron and steel.

Angle of repose: The maximum acute angle that the inclined surface of a pile of loosely divided material naturally makes with the horizontal.

ANSI: American National Standards Institute.

APC: Air Pollution Control.

Apron conveyor: A set of continuous chains that are supported by a system of sprockets and rollers while carrying overlapping or interlocking plates upon which bulk materials are moved.

Ash: The inert residue that remains after a solid waste and fuel mixture has been incinerated.

Baler: A machine used to compress recyclables into bundles to reduce volume. Balers are often used on newspaper, plastics, and corrugated cardboard.

Ballistic separator: A device that drops mixed materials having different physical characteristics onto a high-speed rotary impeller; they are hurled off at different velocities and land in separate bins.

Baffle: A construction used to close or deflect the delivery of a moving substance.

Bond paper: Term originally meant paper used for printing bonds and stocks, now generally refers to high grade papers used for letters and high quality printed work. It is surface-sized for better writing and printing quality.

Bridge crane: A lifting unit that can maneuver horizontally in two directions.

Briquetter: A machine that compresses a material, such as metal turnings, coal dust, or RDF (refuse derived fuel), into objects, usually shaped like a pill, pellet, or pillow.

Broker: An individual or group of individuals that acts as an agent or intermediary between the sellers and buyers of recyclable materials.

Buffer zone: Neutral area which acts as a protective barrier separating two conflicting forces. An area which acts to minimize the impact of pollutants on the environment or public welfare. For example, a buffer zone is established between a composting facility and neighboring residents to minimize odor problems.

Bulk density: The weight in air of a volume of material including voids normal to the material.

Bulky waste: Large items of refuse including, but not limited to, appliances, furniture; large auto parts; nonhazardous construction and demolition materials; and trees, branches, and stumps which cannot be handled by normal solid waste processing, collection, and disposal methods.

Buy-back center: A facility where individuals bring recyclables in exchange for payment.

By-products: Materials which result from operation of a facility and which cannot be composted; but which can, within reason, be recycled, marketed, processed, or otherwise utilized.

Capacity factor: The ratio of the average load on a machine or equipment for the period of time considered, to the capacity rating of the machine or equipment.

Clamshell bucket: A vessel used with a hoist to convey materials; it has two jaws that clamp together when the vessel is lifted by specially attached cables.

Clean Air Act: Act passed by Congress to have the air "safe enough to protect the public's health." Requires the setting of National Ambient Air Quality Standards (NAAQS) for major primary air pollutants.

Clean Water Act: Act passed by Congress to protect the nation's water resources. Requires the EPA to establish a system of national effluent standards for major water pollutants, requires all municipalities to use secondary sewage treatment, sets interim goals of making all U.S. waters safe for fishing and swimming, allows point source discharges of pollutants into waterways only with a permit from the EPA, requires all industries to use the best practicable technology (BPT) for control of conventional and nonconventional pollutants, and to use the best available technology (BAT) that is reasonable or affordable.

Commercial waste: Waste materials originating in wholesale, retail, institutional, or service establishments such as office buildings, stores, markets, theaters, hotels, and warehouses.

Commingled recyclables: A mixture of several recyclable materials.

Comminution: Size reduction.

Compaction: Compressing wastes to reduce their volume. Compaction allows for more efficient transport, but may reduce aeration.

Compactor: Power-driven device used to compress materials to a smaller volume.

Computer printout paper: Consists of white sulfite or sulfate papers in forms manufactured for use in data processing machines. This grade may contain colored stripes and/or computer printing, and may contain not more than 5% of groundwood in the packing. A stock must be untreated and uncoated.

Container deposit legislation: Laws that require monetary deposits to be levied on beverage containers. The money is returned to the consumer when the containers are returned to the retailer. Also called "Bottle Bills."

Contaminant: Undesirable constituent.

Corrugated paper: Paper or cardboard manufactured in a series of wrinkles or folds, or into alternating ridges and grooves.

Cullet: Clean, generally color-sorted, crushed glass used to make new glass products.

Curbside collection: Programs where recyclable materials are collected at the curb, often from special containers, to be brought to various processing facilities.

Cyclone separator: A cylindrical and conical structure without moving parts, which utilizes centrifugal force to remove solids entrained in an air stream.

Dense media separation: A separation process of nonferrous metal from other large particles such as rubber, plastic, bone, or leather, using a fluid solution with a specific gravity about twice that of water. The metal fraction sinks in the solution while other material floats.

Densification equipment: Balers, pellet mills, briquetters, cubetters, etc.

Densified refuse-derived fuel (d-RDF): A refuse-derived fuel that has been processed to produce briquettes, pellets, or cubes.

Density: The mass divided by the volume at a stated temperature.

Design capacity: The quantity of material that a designer anticipates his system will be able to process in a specified time period under specified conditions.

Detinning: Recovering tin from "tin" cans by a chemical process which makes the remaining steel more easily recycled.

Diversion rate: A measure of the amount of waste materials being diverted for recycling compared with the total amount that was previously discarded.

Drag conveyor: A conveyor that uses a series of mechanical barriers such as steel bars fastened between two continuous chains to drag material along a smooth surface.

Drop-off center: A method of collecting recyclable or compostable materials in which the materials are taken by individuals to collection sites and deposited into designated containers.

Dry process: Processes which handle or process solid waste directly as received without the addition of water.

Dust: A loose term applied to solid particles predominantly larger than colloidal and capable of temporary suspension in air or other gases.

Dusts do not tend to flocculate except under electrostatic forces; they do not diffuse but settle under the influence of gravity.

Dust loading: An engineering term for "dust concentration"--among others, usually applied to the contents of air or gas ducts and emissions from stacks, expressed in grains per cubic foot or pounds per thousand pounds of gas or other equivalent units.

Eddy current separator: A device which passes a varying magnetic field through feed material, thereby inducing eddy currents in the nonferrous metals present in the feed. The eddy currents counteract the magnetic field and exert a repelling force on the metals, separating them from the field and the remainder of the feed.

Effluent: Any solid, liquid, or gas which enters the environment as a by-product of a man-oriented process. The substances that flow out of a designated source.

Electronic-optical sorter: Separates glass from stones and pieces of ceramics; sorts the glass according to color. Photoelectric detector determines the color or opacity of the material and blasts of air deflect the pieces into the proper containers.

Electrostatic precipitator: Device for removing particulate matter from MWC facility air emissions. It works by causing the particles to become electrostatically charged and then attracting them to an oppositely charged plate, where they are precipitated out of the air.

Emission: Discharge of a gas into atmospheric circulation.

Energy recovery: Conversion of waste energy, generally through the combustion of processed or raw refuse to produce steam. See also Municipal Waste Combustion and Incineration.

Enterprise fund: A fund for a specific purpose that is self-supporting from the revenue it generates.

EPA: (United States) Environmental Protection Agency.

Feedstock: Waste material furnished to a machine or process.

Ferrous metals: Predominantly iron and steel materials (typically contains small amounts of paper, textiles, plastic, and nonferrous metals) - can be recovered by magnetic separation.

Fines: Very short pulp fibers or fiber fragments escape during paper forming in the process water; may be recovered for reuse or go into sludges. Waste paper processing creates extensive fines.

Finished products Wood chips, manure, screened compost, and other products produced from Acceptable Yard Debris.

Firm capacity: Assumed facility processing capacity accounting for equipment vulnerability.

Flail: A metal flange or tine attached to a rotating shaft for moving, mixing, and aerating leaves.

Flat glass: A general term covering sheet glass, plate glass, and various forms of rolled glass.

Flight conveyor: A drag conveyor that has rollers interspersed in its pull chains to reduce friction.

Flint glass: A lead-containing colorless glass.

Flow control: A legal or economic means by which waste is directed to particular destinations. For example, an ordinance requiring that certain wastes be sent to a combustion facility is waste flow control.

Fly ash: Small, solid particles of ash and soot generated when coal, oil, or waste materials are burned. Fly ash is suspended in the flue gas after combustion and is removed by the pollution control equipment.

Front-end loader: A tractor vehicle with a bucket-type loader at the front end of the vehicle.

Front-end recovery: Mechanical processing of as-discarded solid wastes into separate constituents.

Froth flotation: A process for separating, in aqueous suspension, finely divided particles that have different surface characteristics. Reagents are selected which, when added to the mixture, will coat only the desired material and make their surfaces water-repellent (hydrophobic). When air is bubbled through the solution, the coated particles become affixed to the air bubbles and are buoyed to the surface where they can be removed as froth.

Grade: A term applied to a paper or pulp which is ranked (or distinguished from other papers or pulps) on the basis of its use, appearance, quality, manufacturing history, raw materials, performance, or a combination of these factors.

Gravity separation: Concentration or separation of a mix of materials based on differences in specific gravity and sizes of materials.

Ground wood pulp: A wood pulp produced mechanically by a grinding action that separates wood fibers from resinous binders. It is used principally for newsprint and printing papers.

Hammermill: A type of crusher or shredder used to break up waste materials into smaller pieces.

HDPE: High-density polyethylene containers (containers for milk, liquid detergents, bleach, film, cosmetics, and medicines).

Heavy media separation: Separation of solids into heavy and light fractions in a fluid medium whose density lies between the fractions.

Heavy metals: Dense metals, specifically cadmium, mercury, lead, copper, silver, zinc, and chromium, which may be found in the waste stream. High concentrations in compost can restrict use.

HHW: Household hazardous waste.

High-grade paper: Relatively valuable types of paper such as computer printout, white ledger, and tab cards. Also used to refer to industrial trimmings at paper mills that are recycled.

Horsepower, shaft (flywheel or belt horsepower): Actual horsepower produced by an engine, after deducting the drag of accessories.

Inclined plate conveyor: A separating device that operates by feeding material onto an inclined steel plate backed belt conveyor so that heavy and resilient materials, such as glass, bounce down the conveyor, and light and inelastic materials are carried upward by the motion of the belt.

Inertial separator: Device that relies on ballistic or gravity separation of materials having different physical characteristics.

Inorganic waste: Waste composed of matter other than plant or animal (i.e., contains no carbon).

Institutional waste: Waste materials originating in schools, hospitals, prisons, research institutions, and other public buildings.

Integrated solid waste management: A practice of using several alternative waste management techniques to manage and dispose of specific components of the municipal solid waste stream. Waste management alternatives include source reduction, recycling, composting, energy recovery, and landfilling.

Intermediate processing center (IPC): Usually refers to the type of MRF that processes residentially collected mixed recyclables into new products available for market; often used interchangeably with MRF.

IPC: See intermediate processing center.

IRB: Industrial revenue bond.

Kraft paper: A paper made predominantly from wood pulp produced by a modified sulfate pulping process. It is a comparatively coarse paper particularly noted for its strength, and in unbleached grades is used primarily as a wrapper or packaging material.

LDPE: Low-density polyethylene containers (trash bags, diaper backing, fruit and vegetable self-serve bags, storage bags).

Legnin: An amorphous polymeric substance related to cellulose that, together with cellulose, forms the woody cell walls of plants and the cementing material between them.

Live bottom bin: A storage bin for shredded or granular material whereby controlled discharge is by a mechanical or vibrating device cross the bin bottom.

Live bottom pit: A storage pit, usually rectangular, receiving truck unloaded material, utilizing a push platen or bulkhead, reciprocating rams or mechanical conveyor across the pit floor for controlled discharge (retrieval) of the material.

Magnetic fraction: The portion of municipal ferrous scrap remaining after the nonmagnetic contaminants have been manually removed and the magnetic fraction washed with water and dried at ambient temperature or as required by ASTM C29.

Magnetic separation: A system to remove ferrous metals from other materials in a mixed municipal waste stream. Magnets are used to attract the ferrous metals.

Magnetic separator: A device available in several forms, used to remove iron and steel from a stream of material.

Mandatory recycling: Programs which by law require consumers to separate trash so that some or all recyclable materials are not burned or dumped in landfills.

Manual separation: The separation of recyclable or compostable materials from waste by hand sorting.

Mass burn: Combustion of solid waste without preprocessing, as in a mass burn incinerator.

Material balance: An accounting of the weights of material entering and leaving a process usually made on a time related basis.

Material specification: Stipulates the character of certain materials to meet the necessary performance requirements.

Mechanical collector: A device that separates entrained dust from gas through the application of inertial and gravitational forces.

Mechanical separation: The separation of waste into various components using mechanical means, such as cyclones, trommels, and vibrating screens.

Mixed MSW: MSW that has not undergone source separation.

Mixed paper: Low-grade recyclable paper (paperboard, books, catalogs, construction paper, glossy coated paper (except magazines).

Monorail crane: A lifting unit, suspended from a single rail, that can only move in one horizontal direction.

MRF: Materials Recovery Facility.

MSW: Municipal Solid Waste.

NAAQS: National Ambient Air Quality Standards.

NESHAP: National Emission Standards for Hazardous Air Pollutants.

New corrugated cuttings: Consists of baled corrugated cuttings having two or more liners of either jute or Kraft. Nonsoluble adhesives, butt rolls, slabbed or hogged medium, and treated medium or liners are not acceptable in this grade.

Newsprint: A generic term used to describe paper of the type generally used in the publication of newspapers. The furnish is largely mechanical wood pulp, along with some chemical wood pulp.

NIMBY: Acronym of "Not In My Back Yard" - expression of resident opposition to the siting of a solid waste facility based on the particular location proposed.

Noncompostable: Incapable of decomposing naturally or of yielding safe, nontoxic end products. Noncompostable materials include glass, batteries, cans, etc.

Nonferrous metal: Any metal other than iron and its alloys.

NRC: National Recycling Coalition; now called RAC (Recycling Advisory Council).

NRHW: Nonregulated hazardous waste.

NSPS: New Source Performance Standards - EPA's rule which requires the removal of 25% of the waste stream as the best available control technology (BACT) for WTE plants.

NSWMA: National Solid Waste Management Association.

OBW: Oversize bulky waste.

OCC: Old corrugated cardboard.

ONP: Old newspapers.

Organic waste: Waste material containing carbon-to-carbon bonds and being biodegradable. The organic fraction of municipal solid waste includes paper, wood, food wastes, plastics, and yard wastes.

Particle: A small, discrete mass of solid or liquid matter, including aerosols, dusts, fumes, mists, smokes, and sprays.

Particle size: An expression of the size of liquid or solid particles expressed as the average or equivalent diameter or minimum of two linear dimensions.

Performance bond: A bond or other instrument guaranteeing the performance of all obligations of the proposer or guarantor to acquire and construct a facility.

Performance specification: States the desired operation or function of a product or process but does not specify the materials from which the product must be constructed.

Performance test: A test devised to permit rigorous observation and measurement of the performance of a unit of equipment or a system under prescribed operating conditions.

PET: Polyethylene terepthalate (carbonated soft drink bottles) (beverage containers redeemable under the California bottle bill, AB 2020).

Picking table or belt: Table or belt on which solid waste is manually sorted and certain items are removed. Normally used in composting and materials salvage operations.

Post-consumer recycling: The reuse of materials generated from residential and commercial waste, excluding recycling of materials from industrial processes that has not reached the consumer, such as glass broken in the manufacturing process.

Post-consumer waste: Material or product that has served its intended use and has been discarded for disposal after passing through the hands of a final user. Part of the broader category, "recycled material."

PP: Polypropylene (syrup bottles, yogurt and margarine tubs, shampoo containers, container caps and lids, drinking straws).

Primary materials: Virgin or new materials used for manufacturing basic products. Examples include wood pulp, iron ore, and silica sand.

PS: Polystyrene (disposable dishes, cups, bowls, egg cartons).

PSD: Prevention of significant deterioration.

Pulverization: The crushing or grinding of materials into very fine particle size.

PVC: Polyvinyl chloride (meat wrap, bottles for edible oils).

RAC: Recycling Advisory Council; formerly NRC (National Recycling Coalition).

Rated capacity: The quantity of material that the system can process under demonstrated test conditions.

RCRA: Resource Conservation and Recovery Act.

Recovery: The process of retrieving materials or energy resources from wastes. Also referred to as extraction, reclamation, recycling, and salvage.

Recyclables: Materials that still have useful physical or chemical properties after serving their original purpose and that can, therefore, be reused or remanufactured into additional products.

Recycling: The process by which materials otherwise destined for disposal are collected, reprocessed or remanufactured, and reused.

Refuse: Putrescible and nonputrescible solid wastes, except body wastes, and including kitchen discards, rubbish, ashes, incinerator ash, incinerator residue, street cleanings, and market, commercial, office, and industrial wastes.

Refuse Derived Fuel (RDF): Boiler fuel made by shredding and screening solid waste into a material of relatively uniform handling and combustion properties. Often, recyclables can be recovered from the RDF process.

Residential waste: Waste materials generated in single and multiple-family homes.

Residue: Materials remaining after processing, incineration, composting, or recycling have been completed. Residues are usually disposed of in landfills.

Residue conveyor: A conveyor, usually of the drag or flight type, used to remove incinerator residue from a quench trough to a discharge point.

Resource recovery: A term describing the extraction and utilization of materials and energy from the waste stream. The term is sometimes used synonymously with energy recovery.

Reuse: The use of a product more than once in its same form for the same purpose; e.g., a soft drink bottle is reused when it is returned to the bottling company for refilling.

RFP: Request For Proposal.

Roll-off container: A large waste container that fits onto a tractor trailer that can be dropped off and picked up hydraulically.

Rotary screen: An inclined meshed cylinder that rotates on its axis and screens material places in its upper end. Also known as trommel.

Scavenger: One who removes materials at any point in the solid waste management system.

Scrap: Discarded or rejected industrial waste material often suitable for recycling.

Screen: A surface provided with apertures of uniform size. A machine provided with one or more screening surfaces to separate materials by size.

Screening: The process of passing compost through a screen or sieve to remove large organic or inorganic materials and improve the consistency and quality of the end-product.

Screw conveyor: A rotating shaft with a continuous helical flight to move granular type material, along a trough or tube.

Secondary material: A material that is used in place of a primary or raw material in manufacturing a product.

Shear shredder: A size reduction machine that cuts material between large blades or between a blade and a stationary edge. See Grinder, Hammermill, Shredder.

Shredder: A mechanical device used to break up waste materials into smaller pieces, usually in the form of irregularly shaped strips. Shredding devices include tub mill grinders, hammermills, flail mills, shears, drum pulverizers, wet pulpers, and rasp mills.

Size-reduction equipment: Hammermills, shredders, etc.

Solid waste: Garbage, refuse, sludge from a water treatment plant or air contaminant treatment facility, and other discarded waste materials and sludges in solid, semi-solid, liquid, or contained gaseous form, resulting from industrial, commercial, mining and agricultural operations, and from community activities.

Source reduction: The design, manufacture, acquisition, and reuse of materials so as to minimize the quantity and/or toxicity of waste produced. Source reduction prevents waste either by redesigning products or by otherwise changing societal patterns of consumption, use, and waste generation.

Source separation: The segregation of specific materials at the point of generation for separate collection. Residences source separate recyclables as part of a curbside recycling program.

Special waste: Refers to items that require special or separate handling, such as household hazardous wastes, bulky wastes, tires, and used oil.

Stack emissions: Air emissions from a combustion facility's stacks.

Storage: The interim containment of solid waste, in an approved manner, after generation and prior to ultimate disposal. See live bottom bin.

Storage pit: A pit in which solid waste is held prior to processing.

Subtitle C: The hazardous waste section of the Resource Conservation and Recovery Act (RCRA).

Subtitle D: The solid, nonhazardous waste section of the Resource Conservation and Recovery Act (RCRA).

Subtitle F: Section of the Resource Conservation and Recovery Act (RCRA) requiring the federal government to actively participate in procurement programs fostering the recovery and use of recycled materials and energy.

SWDA: Solid Waste Disposal Act.

SWMP: Solid waste management plan.

Tin can: A container made from tin-plated steel

Tipping fee: A fee, usually dollars per ton, for the unloading or dumping of waste at a landfill, transfer station, recycling center, or waste-to-energy facility, usually stated in dollars per ton; also called a disposal or service fee.

Tipping floor: Unloading area for vehicles that are delivering municipal solid waste to a transfer station, recycling center, composting facility, or municipal waste combustion facility.

TPD: Tons per day.

TPH: Tons per hour.

TPW: Tons per week.

TPY: Tons per year.

Transfer station: A permanent area where waste materials are taken from smaller collection vehicles and placed in larger vehicles for transport, including truck trailers, railroad cars, or barges. Recycling and some processing may also take place at transfer stations.

Trash: Material considered worthless, unnecessary, or offensive that is usually thrown away. Generally defined as dry waste material, but in common usage it is a synonym for garbage, rubbish, or refuse.

Trommel: A perforated rotating essentially horizontal cylinder (a hollow cylindrical screen) used to break open trash bags, screen large pieces of glass and remove small abrasive items such as stones and dirt.

Variable container rate: A charge for solid waste services based on the volume of waste generated measured by the number of containers set out for collection.

Vibrating screen: An inclined screen that is vibrated mechanically, and screens material placed on it.

Virgin materials: Material derived from substances mined, grown, or extracted from water or the atmosphere, and virgin materials are juxtaposed to secondary materials.

VOC: Volatile organic compounds.

Volume reduction: The processing of waste materials so as to decrease the amount of space the materials occupy, usually by compacting or shredding (mechanical), incineration (thermal), or composting (biological).

Waste exchange: A computer and catalog network that redirects waste materials back into the manufacturing or reuse process by matching companies generating specific wastes with companies that use those wastes as manufacturing inputs.

Waste paper: Any paper or paper product which has lost its value for its original purpose and has been discarded.   The term is most commonly used to designate paper suitable for recycling, as paper stock. Paper waste generated in the paper manufacturing process itself is excluded.

Waste reduction: Reducing the amount or type of waste generated. Sometimes used synonymously with Source Reduction.

Waste stream: A term describing the total flow of solid waste from homes, businesses, institutions, and manufacturing plants that must be recycled, burned, or disposed of in landfills; or any segment thereof, such as the "residential waste stream" or the "recyclable waste stream."

WDF: Waste derived fuel facility.

Wet ton: Two thousand pounds of material, "as is." It is the sum of the dry weight of the material, plus its moisture content. Yard waste weighed on truck scales would typically be reported this way.

White goods: Large household appliances such as refrigerators, stoves, air conditioners, and washing machines.

WTE: Waste-to-energy.

Yield: The quantity or percentage of recovered product(s) from the process.

APPENDIX B

List of Material Grades and Specifications

Table 1. Examples of Buyer Specifications for Newspaper.

Buyer	Baled	Loose	Bundled	Grade	Rotogravure	Colored	OCC	Grocery Bags	Maximum Accepted	Delivery
					Contamination					
A	X	X	–	#7	Normal	Normal	–	–	160 TPM	Trailer
B	X	X	–	#7	–	–	–	–	No limit	Self-dump
C	NO	X	X	–	No glossy	–	None	X	20-40 TPW	Semi-trailer
D	NO	X	X	#6	–	–	–	X	No limit	Self-dump
E	X	–	–	–	–	–	–	–	–	Truck/Rail
F	X	–	–	–	–	–	–	–	–	Flatbed/Van

X = Acceptable
– = Not specified

Table 2. Examples of Buyer Specifications for OCC

Buyer	Baled	Loose	Maximum Quantity Accepted	Method of Delivery
A	X	–	–	Truck/rail
B	X	X	No limit	Self-dump
C	X	–	–	Flatbed/van

X = Acceptable
– = Not specified

Table 3. Examples of Buyer Specifications for Tin Cans

Buyer	Baled	Briquet	Loose	Flattened	Contamination W/Bi-Metal	Contamination Food & Labels	Delivery
A	70 lbs/cu ft	70 lbs/cu ft	–	–	X	X	–
B	75 lbs/cu ft	–	–	–	X	NO	Truck or Rail
C	50 lbs/cu ft	–	X	–	–	X	Truck
D	X	–	X	NO	–	–	Flatbed/Van

X = Acceptable
– = Not specified

GENERAL INFORMATION[3]

a. Cleanness. All grades shall be free of dirt, nonferrous metals, or foreign material of any kind, and excessive rust and corrosion. However, the terms "free of dirt, nonferrous metals, or foreign material of any kind" are not intended to preclude the accidental inclusion of negligible amounts where it can be shown that this amount is unavoidable in the customary preparation and handling of the particular grade involved.

b. Off-grade material. The inclusion in a shipment of a particular grade of iron and steel scrap of a negligible amount of metallic material which exceeds to a minor extent to meet the applicable requirements as to qualify or kind of material, shall not change the classification of the shipment provided it can be shown that the inclusion of such off-grade material is unavoidable in the customary preparation and handling of the grade involved.

ISRI code number	Selected Definitions:
209[B]	**No. 2 bundles.** Old black and galvanized steel sheet scrap, hydraulically compressed to charging box size and weighing not less than 75 lbs per cu ft. May not include tin or lead-coated material or vitreous enameled material.
211	**Shredded Scrap.** Homogeneous iron and steel scrap magnetically separated, originating from automobiles, unprepared No. 1 and No. 2 steel, miscellaneous baling and sheet scrap. Average density 70 lb/cu ft.
213	**Shredded Tin Cans for Remelting.** Shredded steel cans, tin-coated or tin-free, may include aluminum tops but must be free of aluminum cans, nonferrous metals except those used in can construction, and non-metallics of any kind.
215	**Incinerator bundles.** Tin can scrap, compressed to charging box size and weighing not less than 75 lbs/cu ft. Processed through a recognized garbage incinerator.

[a]Adapted from Scrap Specifications Circular 1990, Institute of Scrap Recycling Industries, Inc. (ISRI).
[b]Current price often used as a basis by buyers for establishing price for tin cans.

Table 4. Example of Specifications for aluminum Used Beverage Containers (UBC)[1]

	Shredded	Densified	Baled	Actual Buyer Specifications Baled
Density lbs/cu ft	12-17	35-45	14-17 unflattened 22 flattened	14 to 24 - 30
Size	–	Uniform for a bundle 10" to 13" x 10¼" to 20" x 6¼" to 9" Bundle: 41" to 44" x 51" to 54" x 54" to 56" height	30 cu ft minimum. 24" to 40" x 30" to 52" x 40" to 84"	30 cu ft minimum 24" to 40" x 30" to 52" x 40" to 72"
Ferrous Separation	Magnetic	Magnetic	Magnetic	–
Free Lead	None	None	None	–
Steel, lead, bottle caps, plastic, cans, other plastics, glass, wood, dirt, grease, trash, and other foreign substances	None	None	None	–
Tying Method	–	4 to 6 5/8" x 0.020" steel bands or 6 to 10 #13 ga steel wires (or aluminum bands or wires of equivalent strength and number).	4 to 6 5/8" x 0.020" steel bands or 6 to 10 #13 ga steel wires (or aluminum bands or wires of equivalent strength and number).	A minimum of 6 5/8" x 0.020 steel straps or 6 to 10 #13 ga steel wires or equivalent are required. Aluminum bands or wires are acceptable in equivalent strength and number. Bands or wire of other material are not acceptable.
Skids and/or support sheets	–	Not acceptable	Not acceptable	Support sheets are not acceptable
Aluminum items other than UBC	Not acceptable	Not acceptable	Not acceptable	–
Other items	Including moisture by special arrangement between buyer and seller	Including moisture, by special arrangement between buyer and seller	By special arrangement between buyer and seller	Composite bales of two or more individual bales banded together to meet size specifications are not acceptable.

Table 4. Example of Specifications for aluminum Used Beverage Containers (UBC)[1]

(CONTINUED)

Other Conditions	Max. of 5% fines less than 4 mesh. Max. of 2.5% fines less than 12 mesh.	Biscuit shall have banding slots in both directions to facilitate banding. One vertical band per row and minimum of two horizontal bands per bundle.	–	–

NOTE: Individual buyers' specifications may differ. Some buyers will accept (and may prefer) UBCs flattened and pneumatically conveyed to transport trailers. When buyer provides flattener/blower at no cost, often a guaranteed monthly volume (e.g., 25,000 lbs) is required.

– = Not specified.

---

[1] Adapted from Scrap Specifications Circular 1990. Institute of Scrap Recycling Industries, Inc.

Table 5. Examples of Buyer Specifications for PET.

Buyer	Baled	Granulate	Clear	Green	Contamination						Delivery
					Caps	Labels	Ferrous	HDPE	PVC	Other	
A	15 lb/cu ft	No	–	–	X	X	–	–	No	2% maximum other plastics, metals, paper	Truckload
B	3'x4'x5', 10 lb/cu ft	No	–	–	X	X	–	Mix ok	No		25-30,000 lb trailer
C	Color sort	X .	X	X	–	–	–	–	–	–	Granulated in gaylords
D	Maximum density	X	X	X	–	–	–	–	–	3% maximum contamination	Granulated in gaylords
E	X	–	–	–	–	–	–	–	–	–	–
F	–	X	–	–	–	No	No	–	–	No bottle bottoms	2,000 lb super-sack

X = Acceptable
– = Not specified
PVC = Polyvinyl Chloride

Table 6. Examples of Buyer Specifications for HDPE.

Buyer	Milk	Non-Milk	Baled	Granulate	Contamination								Delivery
					Caps	Ferrous	PP	PET	PVC	Moisture	UV Degraded	
A	X	–	Separate or Mix (a)	No	X	–	–	Soda	No	–	–	25,000 lb trailer load
B	X	X	Separate	No	No	No	–	–	–	–	–	30-40,000 lb trailer load
C	X	X	Separate	X	–	–	No	No	No	Low	Low	Granulated in gaylords
D	–	–	700-800 lb/bale	X	No	–	–	–	–	–	–	
E	–	–	No	X	–	–	–	–	–	–	–	
F	X	X	X	No	–	–	–	–	–	–	–	36,000 lb trailer load

X = Acceptable
– = Not specified
PP = Polypropylene
PVC = Polyvinyl Chloride
UV = Ultraviolet
(a) with PET

Table 7. Examples of Buyer Specifications for Collet (Mid-West)

| Buyer | Clear | Green | Amber | Mixed | Crushed | Contamination | | | Delivery |
						Metal	Other	Moisture	
A	X	X	–	–	1/2" minimum	None	Labels ok	Dry	Rear Dump or Rail
B	X	–	X	–	1/2" minimum	None	Labels ok	dry	Rear Dump
C	X	–	X	–	No	–	–	–	Rear Dump
D	X	–	X	–	1/4" - 2" or whole	None	–	1/2% maximum	Self Dump
E	X	X	X	X	1/4" - 2" or whole	None	No Plastic	–	Self Dump or Rail

a) Generally unacceptable materials include:
   Non-container glass (window, pyrex, lab materials, light bulbs, etc.)
   Metal shavings or metal pieces

X = Acceptable
– = Not specified

Table 8. Examples of Buyer Specifications for Unprocessed Collet (West Coast)

Buyer	Clear	Green	Amber	2 Color Mix	3 Color Mix	Ferrous pcs > 8x8x12	Ferrous pcs > 1/2" of < 8x8x12	Ferrous pcs < 1/2"	Contamination + 3/4" glass pkg mat'l (b)	Non-ferrous Metal + 3/4" non-glass pkg mat'l (lead, copper, brass)	Non-ferrous Metal -3/4" to + 8 mesh	Non-ferrous Metal -8 to + 20 mesh	Refractory Material (a) +8 mesh	Refractory Material (a) -8 to +20 mesh	Refractory Material (a) -20 to +40 mesh	Organic Materials (labels, etc.) (d)	Cullet Sizing < 3/4"	Delivery
A	95-100%	80-100%	90-100%	90-100% A/G	80-100% A/G	0	1% max	0.05%	normal amts	0-5% max	–	–	0	1	40	normal amts 0.5%	25% max	self-dump vehicles
	0-3% Amber	0-15% Green	0-10% Green															
	0-1% Green	0-5% Flint	0-5% Flint	0-10% Flint	0-20% Flint													
	0-1% other		0-5% Other	Flint	Flint													
B	95%	90%	90%	–	–	0	1% max	0.05%	normal amts	0.5%	0	1	0	1	40	normal amts 0.5%	25% max	

a) Ceramics, pottery, etc.
b) Closures, Aluminum foil
c) Glass packaging material (labels, plasti-shields)
d) Non-glass packaging material (paper, wood, rubber)

Loads will be rejected for Other Contaminants which include:
Excessive amounts of dirt, gravel, asphalt, concrete, limestone, garbage, etc., or
Excessive amounts of moisture, or
Contamination caused by burning glass containers, or
Pyrex, oven-ready material, plate glass, automobile glass, light bulbs, etc.

X = Acceptable
— = Not specified

Table 9. Select List of Paper Grades and Definitions. (Adapted from Scrap Specifications Circular 1990, Institute of Scrap Recycling Industries, Inc.)

Grade	Name	Description	% Maximum Prohibitive Materials	% Maximum Total Outthrows
1	Mixed Paper	Consists of a mixture of various qualities of paper not limited as to type of packing of fiber content	2.0	10.0
3	Super Mixed Paper	Consists of a baled clean, sorted mixture of various qualities of papers containing less than 10% of groundwood stock, coated or uncoated	0.5	3.0
6	News	Consists of baled newspapers containing less than 5% of other papers	0.5	2.0
7	Special News	Consists of baled sorted, fresh dry newspapers, not sunburned, free from paper other than news, containing not more than the normal percentage of rotogravure and colored sections.	0.0	2.0
8	Special News De-Ink Quality	Consists of baled corrugated containers having liners of either test liner, jute, or kraft	0.0	0.25
11	Corrugated Containers	Consists of baled brown kraft bags free of objectionable liners or contents	1.0	5.0
15	Used Brown Kraft	Consists of papers which are used in forms manufactured for use in data processing machines. This grade may contain a reasonable amount of treated papers	0.0	0.5
25	Groundwood Computer Printout	Consists of papers which are used in forms manufactured for use in data processing machines. This grade may contain a reasonable amount of treated papers	0.0	2.0

CONTINUED ON NEXT PAGE

Table 9. Select List of Paper Grades and Definitions. (Adapted from Scrap Specifications Circular 1990, Institute of Scrap Recycling Industries, Inc.)

Grade	Name	Description	% Maximum Prohibitive Materials	% Maximum Total Outthrows
30	Sorted Colored Ledge (Post-Consumer)	Consists of printed or unprinted sheets, shavings, and cuttings of colored or white sulphite or sulphate ledge, bond, writing, and other papers which have a similar fiber and filler content. This grade must be free of treated, coated, padded or heavily printed stock	0.5	2.0
40	Sorted White Ledger (Post-Consumer)	Consists of printed or unprinted sheets, shavings, guillotined books, quire waste, and cuttings of white sulphite or sulphate ledge, bond, writing paper, and all other papers which have a similar fiber and filler content. This grade must be free of treated, coated, padded, or heavily printed stock	0.5	2.0
42	Computer Printout	Consists of white sulphite or sulphate papers in forms manufactured of use in data processing machines. this grade may contain colored stripes and/or impact or non-impact (e.g. laser) computer printing, and may contain not more than 5% of groundwood in the packing. All stock must be untreated and uncoated	0.0	2.0

Table 9. Miscellaneous Practices and Definitions (CONTINUED)

A. Baling

Each bale must be secured with a sufficient number of bale ties drawn tight to insure a satisfactory delivery.

B. Tare

Sides and headers must be adequate to make a satisfactory delivery of the packing but must not be excessive, nor can they consist of prohibitive materials. The weight of skids or iron cores should be deducted from a gross invoice weight.

C. Moisture Content

All paper must be packed air dry. Where excess moisture is present in the shipment, the buyer has the right to request an adjustment and if a settlement cannot be reached, the buyer has the right to reject the shipment.

D. Outthrows

The term "Outthrows" is defined as "all papers that are so manufactured or treated or are in such a form as to be unsuitable for consumption as the grade specified."*

E. Prohibitive Materials

The term "Prohibitive Materials" is defined as:

a. Any materials which by their presence in a packing of paper stock, in excess of the amount allowed, will make the packaging unusable as the grade specified.

b. Any materials that may be damaging to equipment.

Note: The maximum quantity of "Outthrows" indicated in connection with the following grade definitions is understood to be the TOTAL of "Outthrows" and "Prohibitive materials."

A material can be classified as an "Outthrow" in one grade and as a "Prohibitive Material" in another grade. Carbon paper, for instance, is "UNSUITABLE" In Mixed Paper and is, therefore, classified as an "Outthrow"; whereas it is "UNUSABLE" in White Ledge and in this case classified as a "Prohibitive Manual."

APPENDIX C

Examples of Maintenance Procedures

Example of Maintenance Procedures: Belt Conveyor

Item No.	Description	Frequency
1	Drive	
a	Remove debris from motor cooling fins	W
b	Check gear case oil level	M
c	Check all fasteners and mounting bolts	M
d	Check drive belts for tension and wear	W
e	Replace guard before running	A/R
2	Head and Tail Pulleys	
a	Remove debris	W
b	Lubricate bearings	M
c	Check all fasteners and mounting bolts	M
d	Check take-up for proper belt tension and belt alignment	W
e	Replace guards before running	A/R
3	Idlers	
a	Remove debris	W
b	Lubricate bearings	M
c	Check for frozen idlers	W
d	Check mounting bolts	M
4	Belt	
a	Inspect for damage, wear, and tracking	W
b	Check belt splice	W

Legend: D = Daily; W = Weekly; M = Monthly; A/R = As required.

Example of Maintenance Procedures: Belt Conveyor (CONTINUED)

Item No.	Description	Frequency
5	Skirting	
a	Remove debris	W
b	Check adjustment	W
c	Check for damage	W
d	Check fasteners	M
6	Wipers	
a	Remove debris	W
b	Check adjustment	W
c	Check wear	W
7	Controls	
a	Remove debris	W
b	Check for damage	W
c	Check and adjust emergency shut-off	W
8	Comments	

Legend: D = Daily; W = Weekly; M = Monthly; A/R = As required.

Example of Maintenance Procedures: Magnetic Separator

Item No.	Description	Frequency
1	Drive	
a	Remove debris from motor cooling fins	W
b	Check gear case oil level	M
c	Check all fasteners and mounting bolts	M
d	Check drive belts for tension and wear	W
e	Replace guard before running	A/R
2	Head and Tail Pulleys	
a	Remove debris	W
b	Lubricate bearings	M
c	Check all fasteners and mounting bolts	M
d	Check take-up for proper belt tension and belt alignment	W
e	Replace guards before running	A/R

Legend: D = Daily; W = Weekly; M = Monthly; A/R = As required.

Example of Maintenance Procedures: Magnetic Separator (CONTINUED)

Item No.	Description	Frequency
3	Idlers	
a	Remove debris	W
b	Lubricate bearings	M
c	Check for frozen idlers	W
d	Check mounting bolts	M
4	Belt	
a	Inspect for damage, wear, and tracking	W
b	Check belt splice	W
c	Check wear plates and fasteners	W
5	Magnet	
a	Check oil temperature and oil seepage	M
b	Clean pressure relief valve	M
c	Check oil level	M
d	With magnet off, slack off belt and blow away accumulated tramp iron from magent.	
	Re-tighten belt	W
6	Controls	
a	Remove debris	W
b	Check for damage	W
7	Comments	

Legend: D = Daily; W = Weekly; M = Monthly; A/R = As required.

Example of Maintenance Procedures: Trommel

Item No.		Description	Frequency
1		Drive	
	a	Remove debris from motor cooling fins	W
	b	Check gear case oil level	M
	c	Check all fasteners and mounting bolts	M
	d	Lubricate drive shaft couplings	M
	e	Check drive belts for tension and wear	W
	f	Check universal joints	M
	g	Replace guards before running	A/R
2		Trunnions	
	a	Check trunnion wheels for wear and alignment	M
	b	Lubricate trunnion bearings	M
	c	Check thrust wheels for wear and alignment	M
	d	Lubricate thrust wheels	M
3		Screen	
	a	Remove debris from screen openings	M
	b	Check screen for structural wear or defects	M
4		Controls	
	a	Remove debris	W
	b	Check for damage	W
5		Comments	

Legend: D = Daily; W = Weekly; M = Monthly; A/R = As required.

Example of Maintenance Procedures: Can Flattener

Item No.		Description	Frequency
1		Drives	
	a	Remove debris from motor cooling fins	W
	b	Check gear case oil level	M
	c	Check all fasteners and mounting bolts	M
2		Conveyor	
	a	Remove debris from head and tail shaft pulleys	W
	b	Lubricate bearings	M
	c	Check take-up for proper belt tension and alignment	W
	d	Check belt for wear and damage	W
	e	Check belt splice	W
3		Drum	
	a	Check drum cleats and reverse or replace as required	M
	b	Lubricate bearings	M
4		Blower	
	a	Remove debris from intake and blades	W
	b	Lubricate bearings	M
5		Controls	
	a	Remove debris	W
	b	Check for damage	W
6		Comments	

Legend: D = Daily; W = Weekly; M = Monthly; A/R = As required.

Example of Maintenance Procedures: Baler

Item No.	Description	Frequency
1	**Power Unit**	
a	Remove debris from motor cooling fins	W
b	Check mounting bolts	M
2	**Hydraulic System**	
a	Remove debris from cooler	W
b	Check mounting bolts	M
c	Check hydraulic oil level	M
d	Check for leakage	W
3	**Wire Rolls**	
a	Inspect for quantity and condition	D
4	**Tie System**	
a	Remove debris	D
b	Inspect for damage	D
5	**Shear Knives**	
a	Inspect for sharpening and/or replacement	W
6	**Bale Ejection Chamber**	
a	Remove debris	W
b	Check for damage	W
7	**Controls**	
a	Remove debris	W
b	Inspect for damage	W
c	Check and adjust emergency shut-off	W
8	**Comments**	

Legend: D = Daily; W = Weekly; M = Monthly; A/R = As required.

BIBLIOGRAPHY

Sealy, G.D., "Magnetic Equipment for the Scrap Processing and Recycling Industries," Recycling Today, August, 1976.

Alter, H., S.L. Natof, K.L. Woodruff, W.L. Freyberger, and E.L. Michaels, "Classification and Concentration of Municipal Solid Waste," in Proceedings - Fourth Mineral Waste Utilization Symposium, E. Aleshin (ed.) Bureau of Mines and IIT Research Institute, Chicago, 1974.

Douglas, E. and P.R. Birch, "Recovery of Potentially Reusable Materials from Domestic Refuse by Physical Sorting," Resource Recovery and Conservation, Volume 1, No. 4, 1976.

Twichell, E.S.," Magnetic Separation Equipment for Municipal Refuse," Presented at the 104th Annual American Institute of Mechanical Engineers Meeting, New York City, February 17-19, 1975.

Alter, H., S.L. Natof, K.L. Woodruff, and R.D. Hagen, "The Recovery of Magnetic Metals from Municipal Solid Waste," National Center for Resource Recovery, Inc., November, 1977.

Bendersky, D., D.R. Keyes, M. Luttrell, M. Simister, and D. Viseck, Processing Equipment for Resource Recovery Systems, Volume I - State of the Art, Municipal Environmental Research Laboratory, U.S. Environmental Protection Agency, EPA-600/2-80-007a, Cincinnati, Ohio, July, 1980.

Diaz, L.F., G.M. Savage, and C.G. Golueke, Resource Recovery from Municipal Solid Waste, Volume I - Primary Processing, CRC Press, Boca Raton, Florida, 1982.

Abert, J.G., "Aluminum Recovery - A Status Report", article reprint from N.C.R.R Bulletin , 7(2,3), 1977.

Dalmijn, W.L., W.P.H. Voskuyl, and H.J. Roorda, "Low-Energy Separation of Non-ferrous Metals by Eddy Current Techniques," in Recycling Berlin '79, K.J. Thome-Kozmiensky (ed.) Berlin, Germany, 1979.

Easterbrook, G.E., "Aluminum can't Resist the Power of the Medium," Waste Age, 10(1), 16, 1979.

Bernheisel, J.R., P.M. Bagalman, and W.S. Parker, "Trommel Processing of Municipal Solid Waste Prior to Shredding" in Proceedings 6th Mineral Waste Utilization Symposium, U.S. Bureau of Mines and IIT Research Institute, Chicago, May 2-3, 1978.

Savage, G.M., L.F. Diaz, and G.J. Trezek, "RDF: Quality must precede Quantity," Waste Age, 9(4),100,1978.

Savage, G.M. and G.J. Trezek, "Screening Shredded Municipal Solid Waste," Compost Science, 17(1),7, 1976.

Woodruff, K.L., "Preprocessing of Municipal Solid Waste for Resource Recovery with a Trommel," in Trans. Soc. Min. Eng., 260, 201, 1976.

Hill, R.M. "Rotary Screens for Solid Waste," Waste Age, 18, 33, 1977.

Sullivan, J.F., Screening Technology Handbook, Triple/S Dynamics, Dallas, Texas, 1975.

Murray, D., "Air Classifier Performance and Operating Principles," Presented at the 1978 National Waste Processing Conference, Chicago, May 7-10, 1978.

Murray, D., and C. R. Liddell, "The Dynamics, Operation, and Evaluation of an Air Classifier," Waste Age, 8, 18, 1977.

Chrisman, R.L., "Air Classification in Resource Recovery," National Center for Resource Recovery, Inc., RM 78-1, October, 1978.

Boettcher, B.A., "Air Classification for Reclamation of Solid Wastes," Compost Science, 2(6), 22, 1970.

Fan, D., "On the Air Classified Light Fraction of Shredded Municipal Solid Waste - Composition and Physical Characteristics," Resource Recovery and Conservation, 1, (141), 1975.

Savage, G.M., L.F. Diaz, and G.J. Trezek, "Performance Characteristics of Air Classifiers in Resource Recovery Processing," in Proceedings of the 1980 National Waste Processing Conference, ASME, 1978.

Ham, R.K. and J.J. Reinhardt, Final Report on a Demonstration Project at Madison Wisconsin to Investigate Milling of Solid Wastes Between 1966 and 1972, Volume I, U.S. Environmental Protection Agency, March, 1973.

Marshall, V.C., "Crushing and Grinding -- Critique of Existing Laws," Chemical and Processing Engineering, April, 1966.

Austin, L.G. and R.R. Klimpel, "Theory of Grinding Operations," I and EC Process, Design, and Development, 56:19-29, 1964.

Snow, R.H., "Annual Review of Size Reduction," Power Technology, 5:351-364, 1971-1972.

Bond, F.C., "The Third Theory of Comminution," Trans. AIME, 193:484-494, 1952.

Trezek, G.J. and G.M. Savage, Significance of Size Reduction in Solid Waste Management, EPA-600/2-77-131, Municipal Research Laboratory, Office of Research and Development, U.S. Environmental Protection Agency, Cincinnati, Ohio 45268, July, 1977.

Vesiland, P.A., A.E. Rimer, and W.A. Worrell, "Performance Characteristics of a Vertical Hammermill Shredder," in Proceedings 1980 National Waste Processing Conference, ASME, May, 1980.

Gaudin, A.M. and T.P. Meloy, "Model and a Communition Distribution Equation for Single Fracture," Trans. AIME, 223:40-43, 1962.

Zalosh, R.G., et al, Factory Mutual Research Corporation -- Assessment of Explosion Hazards in Refuse Shredders, prepared for the U.S. Energy Research and Development Administration under Contract No. (49-1)-3737, April, 1976.

Zalosh, R.G. and J.P. Coll, Determination of Explosion Venting Requirements for Municipal Solid Waste Shredders, draft report submitted to the U.S. Environmental Protection Agency, EPA Contract No. 68-03-2880, September, 1981.

Savage, G.M., D.J. Lafrenz, D.B. Jones, and J.C. Glaub, Engineering Design Manual for Solid Waste Size Reduction Equipment, U.S. Environmental Protection Agency, Cincinnati, Ohio, 1982.